DANGER
FROM THE SKIES

The real threat to mankind's existence!

By DAVID BRYANT

DANGER FROM THE SKIES

First paperback edition printed 2015 in the United Kingdom.

A catalogue record for this book is available from the British Library.

ISBN 978-0-957-4944-8-0

Published by

Heathland Books

For more copies of this book, please email: info@spacerocksuk.com

Telephone: 01603 715933

Designed and typeset by Bob Tibbitts ~ (iSET)

Printed in Great Britain

DEDICATION

This book is respectfully dedicated to Luis and Walter Alvarez,
Sir Fred Hoyle and Chandra Wickramasinghe,
all of whom had the courage to make intellectual leaps
against the current of scientific orthodoxy.

To Joyce!

Keep watching
the skies!

David
Bryant

Contents

Acknowledgements

AS always: first and foremost, thanks to my wife Linda, who has never once failed to support and encourage me in everything I do.

Thanks to my big brother, Dr Rob Bryant for persuading me to study Science rather than turning to the dark side of Art and Music!

I have a few good friends who listen attentively to my ideas and comment honestly on them: in no particular order they are Bob Williams, Jason Hughes and Paul Williams.

Thanks also go to my editor and designer Bob Tibbitts for producing such a professional-looking end product.

Foreword

By Dr ROB BRYANT

I'VE known David longer than any other living person and still he never ceases to amaze me. The breadth and depth of his knowledge is truly exceptional. When added to his ability to challenge received wisdom and offer thoughtful, reasoned alternatives, it is really very easy to recommend his books to both general and specialist readers. Indeed, specialist readers are most in need of enlightenment in an age when so many mainstream utterances issue from the so-called "scientific community" (whatever that is). How can we hope to shoot down hypotheses by applying counterfactual evidence, if we must agree with the status quo? Iconoclasm, scepticism and sheer bloody-mindedness have always been the sustaining philosophies of true natural philosophers.

So, I offer a warm welcome to all those who believe that enlightenment is a process, not an end, to read and savour this latest and, in my humble opinion, the best book to be written by my brother. It offers a thought-provoking interpretation of many of the salient facts gathered on the origins of life and of the solar system up to this time. Whether you come to similar conclusions is a complete irrelevance; if reading *"Danger From The Skies"* makes you take a fresh look at one of the most important topics about which mankind has constantly wondered, then David will have achieved what I believe he's set out to accomplish.

Introduction

MY fascination for everything to do with space began in 1957, when I was six years old. Following the death of Soviet dictator Joseph Stalin in 1953, a slight thaw in the Cold War allowed many nations (but not Communist China!) to participate in an unprecedented eighteen month investigation of the Earth and the space around it.

Having been designated International Geophysical Year, this was an incredible opportunity for every branch of Science to make a contribution to our understanding of the planet.

Eleven areas of Earth Science were selected as the focus for research by the sixty-seven participating countries: cosmic and solar radiation, the aurora, gravity, geomagnetism, the ionosphere, meteorology and climate, more precise mapping of longitude and latitude, oceanography, and seismology.

The various IGY expeditions were covered continuously by the media and I still remember following the Fuchs & Hillary Antarctic crossing on my aunt's tiny TV screen and on Pathé News at Saturday morning cinema!

Some of the incredible advances made included confirmation of the existence of tectonic plates and their relative motions, the discovery of the Van Allen radiation belts and the presence of a surprisingly rich fauna on the floors of some of the deepest oceans.

To me, though, the two most memorable occurrences of 1957 were the launch of Sputnik 1, the world's first artificial satellite, and the apparitions of two great comets: Arend-Roland and Mrkos. Without doubt these events set me on the path to an all-too-brief career as a naval pilot and a longer one as a teacher and lecturer of Science.

Despite the huge disappointment of Halley's Comet in 1986, many readers will have enjoyed the sight of a bright comet: I can remember twelve or so, including the beautiful Hale-Bopp in 1995 and West twenty years earlier.

Comet Pan-starrs

Having retired from full-time teaching and lecturing, I have devoted my time to building up a business importing and selling meteorites and writing and lecturing about them. ***Spacerocks UK*** has for several years been the only professional meteorite dealership in the country: this has allowed me to meet and discuss meteoritics with many professional astronomers, gaining from them and others a continuously updated database of new discoveries and theories in the fields of planetary and cometary astronomy.

The author and Rusty Schweickart

Increasingly, I began to form the opinion that my ideas about comets, asteroids and meteoroids and the potential results of their impact upon the Earth differed completely from the 'received wisdom'. This dichotomy was underlined during discussions I had in 2014 with Apollo astronaut and emeritus chairman of the B612 Foundation, Rusty Schweickart. More than anything else, this meeting persuaded me of the existence of a political agenda behind the current predictions of the dangers faced by mankind from asteroidal impacts and the search for technologies to prevent them.

This book is my analysis of past and present cataclysmic impact events, their effect upon our planet and how, in my view, we are being intentionally misled about the nature of the real threat from space!

A final thought: rather than interrupt the flow of the main text, I've added appendices at the end that provide fuller explanations of some of the more obtuse points!

Prologue

IT is early morning: on the fringes of a towering glade of Redwood trees a small herd of Triceratops is browsing on spiny-leaved cycads, stripping away the fronds to get to the stumpy trunks. A large male raises his head and scents the air: something doesn't seem as it should be. He swings his massive bone-frilled head to scan the horizon for potential danger, but the only other dinosaurs he can see are harmless Anatosaurs a few kilometres away. He is about to resume feeding when he notices a spot of brightness appear above a distant range of hills. As he watches, it grows steadily in size until it is larger and brighter than the Sun: silently it disappears behind the hills . . .

Eyes fixed on the horizon, the Triceratops sees a bright flash that lights up the clouds, followed by a curious 'rippling' of the air that rushes towards him and his family group. As this reaches the Anatosaurs, they instantly disappear: a wall of dust and flame advances ever faster: the Triceratops raises his giant head and utters a bellow of alarm: seconds later he, his herd and everything around them are vaporized.

Sixty-five million years ago, a massive impactor smashed into the Yucatan Peninsula, generating heat-pulses and winds that wiped three quarters of the Earth's flora and fauna from the face of the planet.

But what was the object that caused such devastation? And could mankind one day share the dinosaurs' fate?

Chapter 1

THE ORIGINS OF THE EARTH AND SOLAR SYSTEM

WHEN I taught Science in schools, I occasionally reminded my students that, actually, there is no such thing as a fact: rather, we pass examinations by repeating *that which is currently held to be true by those who write the exams!* As an example, theories about the origins of our Universe have changed many times over the years: there's the current **Big Bang** model, the **Steady State Universe**, the **Eternal Inflation Multiverse** and the **Oscillating Universe**. For those with a taste for the *outré*, we have the **Projected Flat Hologram** hypothesis and even the **Digital Simulation** postulate! (Details of all of these are in an appendix at the end of the book!)

These cannot all be correct, of course, so if you're taking your Astronomy & Cosmology finals soon, you might do worse than to find out which theory is currently flavour of the month!

In just the same way, our understanding of the way in which the planets and (of more particular interest, the Earth-Moon system) formed is under continual review and modification: I offer here the model that most people seem to agree upon at the time of writing!

The Universe is presently considered to be just under fourteen billion years old: since the decoupling of matter and energy 300,000 years after the Big Bang, clouds of hydrogen condensed into stars which then sparked into life as hydrogen atoms fused together to form helium. Within these 'solar furnaces', increasingly large atoms were formed by nuclear fusion, generating heat, light and all the other wavelengths of electromagnetic radiation. These massive first generation (Population 111) stars survived just a few million years, before exploding in the unimaginable fury of novae and supernovae. Clouds of gas and heavier elements were projected off into the Cosmos where they themselves gradually condensed to form new stars: this star formation can actually be seen in photographs of 'stellar nurseries' obtained using the Hubble orbiting telescope.

Our own star, the Sun, is itself a second or third generation star, formed around five billion years ago by the collapse and condensation of a solar nebula. The vast majority of the mass in this rotating cloud of dust and gas formed the Sun itself: just **1%** accreted to form the planets, their satellites, the asteroid and Kuyper belts, comets, the Oort Cloud and the interplanetary material that wasn't 'hoovered up' during the accretionary phase.

After the formation of the Sun and its ignition, the first solid objects to condense within the solar nebula were probably small (millimetric) spheres called chondrules. (The solar nebula already contained larger dust particles and pre-solar grains from elsewhere, but chondrules are considered to be the first 'home-grown' mineral particles.

(It should be acknowledged that some researchers are beginning to suggest that some chunks of heavy elements and their compounds such as the nickel-iron minerals taenite and kamacite may also have condensed directly at this time.)

At the moment, nobody is exactly certain how chondrules acquired

Meteorite, polished to show chondrules

their fascinating internal structures or the prime cause for their formation around four and a half billion years ago (there are over sixty current theories in circulation!) but energy from pressure fronts from distant novae has been suggested and seems plausible. What is generally accepted is that chondrules accreted in a kind of 'snowball effect', whereby electrostatic attraction and gravity bound them together into increasingly large lumps: some classes of primitive unequilibrated meteorites seem to display evidence of this process in their structure. Collisions between these created bigger and bigger chunks, ultimately producing *planetissimals*: individual bodies large enough to be considered as 'mini-planets'.

Once these proto-planets had attained sufficient mass, the process of differentiation began: the residual kinetic energies of the collisions that formed the objects as well as radioactive decay and tidal forces from other large bodies nearby would have caused melting and the migration of heavy elements inwards to form a core. This process resulted in the characteristic structure we observe today in the four rocky inner planets: Mercury, Venus, Earth and Mars, as well as in many other large objects in the Solar System such as planetary satellites and asteroids. Most of these have a predominantly nickel-iron core overlain by a rocky mantle and a more or less solid crust.

Lighter elements and their compounds accumulated further from the centre of the solar nebula, eventually condensing to form the outer 'gas giant' planets, Jupiter, Saturn, Uranus and Neptune, all of which exhibit thick atmospheres of low molecular-weight gases such as methane and ammonia.

It seems probable that very low mass water molecules migrated even further, out to the very edge of the solar nebula where they accreted to form cometary nuclei: planet-sized chunks of ice and dust.

Collisions between the protoplanets must have been frequent, until eventually only eight major planets remained: it is widely believed that two of the later collisions may have generated the Earth-Moon system and resulted in the strange structure of the innermost planet, Mercury.

Following the formation of the Solar System as we more or less see it today, many astronomers are confident that at least two periods of bombardment subsequently took place, when large objects crashed onto the newly-formed surfaces of the planets, leaving vast scars that may still be visible. For example, the Moon and many other planetary satellites are pock-marked by huge impact craters: the Moon's more recent maria basins ('seas') were also formed by even more massive impacts. Similar crater-fields have been discovered on Mercury, Venus and Mars: as we shall see, the Earth did not escape, either!

These two periods of cataclysmic impacts are generally referred to as the *Initial* and *Late Heavy Bombardments*: the dates of which are agreed to be between 4.2 and 3.9 billion years ago. The mechanisms that caused these are the subject of conjecture, ranging from the gravitational implications of a later formation date for the two outermost gas giants Uranus and Neptune, to a 'migration' of Jupiter and Saturn, during which these two massive bodies exchanged orbits, with a destabilising effect on the asteroid belt. (The main-belt asteroids are assumed in this model to be the impactors responsible for the LHB)

The generally accepted evidence for the LHB is the heavy cratering of the inner planets, their satellites and of the larger asteroids: the date of the event (around 700 million years after planetary formation was more or less complete) is largely derived from the examination of planetary meteorites. However, physical evidence of the LHB on the surface of the Earth is very patchy and incomplete (for reasons we'll consider in the next chapter!) so it is not possible to be absolutely certain about any of the details.

Remember my comment about scientific facts at the start of this chapter? Here's yet another important part of our understanding of the Earth's history that is also subject to debate and conjecture: how old are the oceans and where did they come from?

Aging of some of the world's oldest minerals (particularly detrital zircons) and sedimentary deposits, and analysis of their relative oxygen isotope concentrations, suggest that the oceans were in place by *either* 3.8 billion *or* 4.2 billion years ago! It is considered axiomatic that life (at least as we know it) is unlikely to evolve or be sustained without liquid water, so the formation of the Earth's hydrosphere (and that of any planet we may suspect of harbouring life!) is critical to our understanding of the mechanisms by which life spontaneously generates in the Universe. (Assuming we believe that life *does* emerge spontaneously, as the end point of some fortuitous chemical reactions!)

Although acceptance of this suggestion ebbs and flows, it seems possible to me that the greater part of the water that now covers our planet derives from cometary impacts.

Most people have a mental image of a comet based around the 'dirty snowball' model popularised by astronomer Fred Whipple: we will return to this topic in Chapter 8. However, for now it is sufficient to note that Halley's Comet (a famous periodic comet that last visited the inner Solar System in 1986) has an icy nucleus measuring roughly 16 x 8 x 8 kilometres! That's a volume of *over one thousand billion cubic metres!* Now a cubic metre of ice has a mass of 916kg: just under a metric tonne: you'll already have concluded that Halley's Comet represents a vast mass of water.

It's not so unlikely (in my estimation, at least) that a steady stream of incoming comets four billion years ago provided ample water to fill the ocean basins. Without any doubt, this bombardment could also have played a significant part in the cooling and stabilisation of the Earth's crust, as cometary ice initially sublimed and then (when surface temperatures dropped below 100°C) evaporated from water into steam. Such changes of state require latent heat: this would have been acquired from the Earth's surface in the same way that sweat cools us down as it evaporates from our skin.

So we have reached an understanding of possible mechanisms by which the early Earth and its oceans were formed, this process being complete by around four billion years ago. But our planet was not – and is not – an immutable lump floating in space: it is a dynamic and active body with an ever-changing surface: we will consider how this has affected our understanding of the Earth's history in the next chapter.

German meteorologist and polar explorer Alfred Wegener

Chapter 2

THE DYNAMIC EARTH

AS noted earlier, one of the major achievements of International Geophysical Year, 1957-58 was proof of the reality of Continental Drift.

As early as the sixteenth century, Flemish cartographer Abraham Ortellius noticed that the western coastline of Africa and that of South America seemed to fit together. He speculated that the two continents might once have actually been joined, but had moved apart at some time in the past. A similar conclusion was reached during the nineteenth century by Alexander von Humboldt and Antonio Snider-Pellegrini, but credit for bringing the theory to a wider audience is generally given to German meteorologist and polar explorer Alfred Wegener. As happens far too frequently, his lack of any formal qualifications in Geology resulted in his thesis being rejected out of hand, but he was vindicated during International Geophysical Year and given the credit he deserved.

Like Ortellius and the others, Wegener suggested that the tessellating coastlines of the South Atlantic were not just the result of a happy coincidence, but proof that the two continents had once formed part of a much larger land mass. Wegener's *original* contribution was the observation that additional evidence came from the fossil record and from existing flora and fauna:

- The fern Glossopteris is found in rocks of Permian age from India, Australia, South Africa, South America and Antarctica. There is no way to explain this except by accepting that these land masses were once conjoined.
- A small Permian reptile, Mesosaurus, is only found in deposits either side of the South Atlantic: as a freshwater species, it could not have swum from West Africa to eastern South America!
- The unique fauna of Australia indicates that it separated from the rest of the continents before the evolution of placental mammals, allowing primitive marsupials and monotremes to continue to develop in isolation.
- Fossils of tropical plants and animals (including coal deposits) are found in Antarctica, necessitating a near-equatorial location for the continent during the Carboniferous Period.

Studying the distribution of such species allows modern geologists to make a very good estimate of when the various continents separated and in

Upthrust caused by colliding plates   **Oceanic crust being subducted**

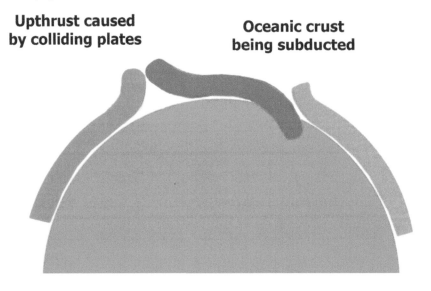

which direction they travelled. As a matter of fact, the current figure for the separation of South America and Africa is around 2.5cm per year.

Wegener had two problems when trying to convince sceptics: firstly, not **all** the continents fit so well together and secondly, he couldn't suggest a mechanism by which vast land-masses could drift about the Earth's surface like croutons on a bowl of onion soup!

Discoveries made during International Geophysical Year produced incontrovertible evidence that the Atlantic Ocean floor is indeed spreading outwards from the centre: it was established that a double ridge of submarine mountains marks a region where basalt is being extruded upwards from the mantle. Dating of rock samples from this mid-Atlantic Ridge revealed that the further they were obtained from the mid-ocean ridge, the older they are.

This is, of course, conclusive evidence that it is the spreading of the Atlantic basin that is the reason for the increasing gulf between South America and West Africa. Once the principle of Continental Drift had been confirmed, further research revealed that similar movements occurred around the globe. It was further established that the Earth's crust is not uniformly thick, but consists of regions of oceanic crust and thicker cratonic (continental) crust. These are split into vast slabs known as Tectonic Plates which may contain **both** oceanic and cratonic crust. These 'float' on the mantle beneath them, hence the more usual term for movements of the Earth's surface: ***Plate Tectonic Drift***. The mechanism that drives the motion of the plates is now fully understood: convection currents within the mantle are responsible for both spreading ocean floors (allowing upwelling of magma to form new oceanic plate) and 'scrunching' existing plates together at their margins, allowing island arcs to form. These eventually accrete to produce new continental crust.

At plate margins, crust can be thrust upwards to form mountain ranges like the Himalayas (which are still rising at around the rate your fingernails grow!) or one plate can be subducted: pushed under another, descending into the mantle where its component rocks and minerals are recycled.

(A *genuine* inconvenient truth for Al Gore, which, I'm sure, won't earn *me* a Nobel Prize: trillions of tonnes of carbon dioxide are released annually at subduction zones by the recycling of carbonate rocks such as limestone and chalk!)

What has Plate Tectonics got to do with the theme of this book? Simply this: the main reason for the sketchy fossil record of the LHB on Earth is that ancient surface rocks over much of the Earth's surface have long ago been recycled. Most of the large craters that may have once existed have been subducted. Additionally, the Earth has an often very active atmosphere that is responsible for several kinds of surface erosion:

Wind erosion

Wind-blown particles wear away even the hardest rocks in time: a bit like the technique of sand blasting that is used to clean rusty metal or old buildings.

Chemical erosion

Carbonate rocks like chalk, limestone and dolomite dissolve in the weak carbonic acid in rain and soil water. (Another huge and uncontrollable source of atmospheric carbon dioxide!)

Thermal stressing

When rocks are warmed by the sun, they expand: when they cool at night, they shrink. Over time, this process will cause even the hardest rocks to crumble.

Freeze-thaw

As most people are aware, when water freezes, it expands: if rainwater in fissures in rocks and rock surfaces freezes in the winter, the resulting pressure is sufficient to shatter the rock.

Water erosion

The force of moving water in rivers and seas may contain phenomenal amounts of kinetic energy: this can batter down cliffs, erode deep river valleys and produce vast quantities of silt that bury land surfaces.

Glaciation

From time to time the polar ice extends southwards and northwards, covering much of both hemispheres of the planet with thick ice sheets: in high ground compacted snow forms glaciers that flow downhill, eroding and pushing aside anything in their path. No-one fully understands the factors that tip the Earth into an Ice Age, but there have been at least five of them: we are either still recovering from the most recent or heading into the next (Another inconvenient truth!)

Taking all these into consideration, it's amazing that *any* large craters exist on Earth! But, as we'll see later, there are a few!

Chapter 3

THE EMERGENCE OF LIFE

OUR planet is around 4.2 billion years old: for nearly half a billion years conditions are considered to have been far too hostile for even the simplest life-forms to exist. Then, about 3.8 billion years ago, the first prokaryotic cells appeared. These are the extremely basic organisms that make up the Bacteria and Archaea. Even today, the vast majority of the Earth's biomass is made up of these simple cells, various types of which can be found in the most inhospitable environments: in soda lakes, at the bottom of the deepest ocean abysses even within the rocks of the Earth's mantle and the boiling water of volcanic springs! It has been demonstrated that viruses, which are even more primitive, can survive the harsh radiation and vacuum of outer space: it is considered possible (particularly by advocates of the theory of panspermia) that planets where life exists are initially populated by viral spores drifting down from the cosmos.

Prokaryote cells frequently clump together to form mats, films or slimes: it is possible that this 'endosymbiosis' may eventually have led to the first eukaryotic cells and, ultimately, to multicellular organisms.

All advanced life-forms on Earth (whether one- or many-celled) exhibit *organisation* at cellular level: that is, their cells contain organelles that carry out various functions. Their genetic material (in the form of the complex DNA molecule) is contained within the nuclear membrane of the cell's nucleus. Other organelles only found in eukaryotic cells include mitochondria that are responsible for the production of cellular energy from sugar and (in the case of green plants) chloroplasts that carry out the process of ***photosynthesis***, by which they combine carbon dioxide and water to produce carbohydrates and oxygen. This process is catalysed by a number of green biomolecules that enable plants to 'trap' and metabolise sunlight.

Photosynthesis probably commenced on Earth around 2.7 billion years ago, when Cyanobacteria evolved that could take in visible light and, using pigments including phycobilins, carotenoids and at least three varieties of chlorophyll, produce oxygen as a metabolic by-product.

(Colonies of very similar, primitive cyanobacteria still exist in the form of large, pillow-like structures called Stromatolites. These are generally found in hypersaline lakes and lagoons in the tropics, although colonies have recently been found at the Giant's Causeway in Northern Ireland!)

This efficient ***autotrophic*** nutrition, the resultant scrubbing of atmospheric carbon dioxide and the increase in the partial pressure of oxygen had a number of important results:

- The temperature of the atmosphere steadily decreased (and hence that of the oceans)
- The ozone layer formed, shielding the Earth's surface from much of the incoming ultra-violet radiation
- Multicellular plants such as algae began to evolve
- Increasing numbers of complex organic compounds such as proteins, fats and carbohydrates were produced
- Complex animal life evolved to exploit this food-source, establishing the ecosystems that we observe today.

It is an astonishing fact that by the end of the Cambrian Period of the Earth's history (around 485 million years ago) *every single phylum* found on Earth today had evolved: no new 'body plan' has appeared since then! This rapid increase in diversity is often referred to as the *Cambrian Explosion*.

Geological Periods

Era	Period	mya
Cenozoic	Quaternary	2.588 - 0
	Neogene	23.03 - 2.588
	Paleogene	66.0 - 23.03
Mesozoic	Cretaceous	145.5 - 66.0
	Jurassic	201.3 - 145.0
	Triassic	252.17 - 201.3
Paleozoic	Permian	298.9 - 252.17
	Carboniferous	358.9 - 298.9
	Devonian	419.2 - 358.9
	Silurian	443.4 - 419.2
	Ordovician	485.4 - 443.4
	Cambrian	541.0 - 485.4

The spontaneous appearance of new forms of animal and plant life is generally agreed to be the result of gene mutation and natural selection. (See Appendix 2) This process is called evolution.

First described by nineteenth-century natural philosophers Wallace, Mendel and Darwin, the concept is often misrepresented, not least by creationists. Often from a background of religious fundamentalism, adherents of this belief system hold that all life on Earth was made by God in a couple of days, as is described in various ancient religious texts. Despite the vast amount of contrary evidence, Christian fundamentalists often claim that the Universe was created around 4004 BC. This date was arrived at in the nineteenth century by Bishop Usher of Armagh, who worked backwards through the Old Testament, calculating how many generations there had been and factoring in the claimed ages of patriarchs such as Moses, Abraham and Methuselah.

Although to most people this kind of unswerving faith seems hard to understand, in many ways it is echoed by proponents of Darwinian Evolution. When Charles Darwin published *On The Origin of Species* in 1859, Geology, Palaeontology and Genetics were in their infancy, and this is obvious when one reads the book today. To Darwin, it must have seemed that fossil hunters and explorers would soon provide additional proof of his theory, but he was over-optimistic about the imminent discovery of key 'missing link' forms. In fact, science began to uncover evidence that the most powerful factors in evolution were not *gradual* changes in climate and landscape but rapid cataclysmic ones. More of this later!

That there were a number of weak areas in Darwinian evolutionary theory was widely acknowledged as early as 1880. From then until the 1930's critics pointed out a number of apparent major flaws:

- Some very ancient species had remained entirely unchanged for hundreds of millions of years (King Crabs, Nautiluses and Crinoids, for example)

- Several examples of apparent 'missing links' were found to have been contemporaries of their evolutionary descendants.

- With the possibly contentious exception of examples such as the Peppered Moth, there is not much evidence that evolution is occurring at the moment.

- There is no evolutionary reason that adequately explains the disappearance of any number of highly successful species.

- Evidence from the fossil record suggests that the rate of appearance of new species often seems to have accelerated during periods when the environment and climate were stable and benign, such as during the Carboniferous Period and Mesozoic Era.

- It is difficult to explain how evolutionary 'halfway stages' of organs and structures could have benefitted an organism or made it 'fitter to survive'. A good example of this is that the three tiny bones of the vertebrate middle ear (the incus, malleus and stapes) are believed to have evolved from the gill rakers of ancestral fish. But at some point they must have ceased to be of any use as part of a respiratory system while not yet providing any advantage as sensory organs.

- Many 'missing link' fossils have so far not been discovered: some of those that have are considered dubious.

- Central to Darwin's work was his study of isolated populations of animals and plants on island groups like the Galapagos: some critics have expressed doubts that the Darwinian model would 'work' on larger, more complex ecosystems.

- Among higher life-forms, mating (and hence procreation) involve an element of choice. Suppose a mutation resulted in a significantly different-looking individual. No matter how 'useful' that difference might become further down the generations, would the owner find a mate prepared to accept it?

- In Darwinian evolution, 'fit' means 'better suited to survive in a given environment' without any implication of improved health, or greater physical robustness. Many hypothetical intermediate forms would have been at a decided physical disadvantage when competing for food or a mate or when protecting themselves and their offspring.

- In terms of Darwin's original theory, it is difficult to account for the strange phenomenon of 'Evolutionary Mimicry'. There are any number of species that have evolved to resemble totally unrelated ones. For example, the Lizard Orchid not only **looks** like a couple of dozen dead lizards impaled on a spiked stick, it smells like it too! How on Earth could evolution – the survival of the fittest – produce this end result? Presumably the benefit to the orchid is that the sight and smell of all this apparent carrion attracts pollinating insects like flies: but how did the long haul to the final phenotype (the observed properties of an organism) work out for the interim stages? Did some have foul-smelling, pretty flowers? Or resemble beautifully-scented dead lizards? (See Appendix 3 for further examples with illustrations.)

This period of re-evaluation between 1880 and 1930 is often referred to as 'The Eclipse of Darwinism': further debate in the 1940s led to the generally accepted modern view that evolution does indeed proceed as a result of adaptation through natural selection, but a nod is given to the earlier theories of George Cuvier, who insisted that abrupt changes were an equally important part of the process. However, it would be accurate to say that the majority of Darwin's ideas are still widely believed and taught in schools and colleges: these central points are:

- New species appear as the sum total of many gradual, small changes over time.

- These changes result from gene mutation: fragments of the DNA code of an organism are altered during faulty replication during gametogenesis (reproductive cell production) or by external factors such as radiation or chemical toxicity.

- If these mutations make the next generation 'fitter' to survive, they will be passed on and may continue to mutate, resulting in a completely new species *that only successfully reproduces with its own kind.*

- Species may become extinct through failure to adapt to changing environments, or because of competition from more recently-evolved, fitter species or by coming into contact with new pathogens (harmful disease organisms).

- The extinction or change in the population of any species in an ecosystem will necessarily affect all the other organisms in that system.

One thing that Darwin certainly neither knew nor theorised about however, is that throughout the history of life on Earth, there have been a number of massive extinction events that erased a significant percentage of the gene pool: it is these catastrophes that provide the missing element in Darwinian evolutionary theory.

Chapter 4

GLOBAL EXTINCTION EVENTS

IF we make the quantum leap to an assumption that the appearance of new species of terrestrial life and the disappearance of former ones are not fully explainable in terms of Charles Darwin's theory as originally published, we must look for amendments to it that take account of discoveries since that time.

It was as recently as 1980 that father and son science team Luis and Walter Alvarez suggested that the extinction of the dinosaurs (and many other contemporary species) was the result of a global catastrophe: specifically, the impact upon the Earth of a huge body from outer space.

That there had been a major extinction event sixty-five million years ago that defined the end of the Cretaceous Period was well known. At that time the dinosaurs abruptly disappeared from the fossil record, following into oblivion many other wonderful creatures, the fossil skeletons and painted representations of which fired many a youthful imagination. (As a matter of fact, it is the case that many people's favourites, the ichthyosaurs and plesiosaurs, had vanished some time before the end of the Cretaceous Period). Additionally, many groups of marine invertebrate that had, until then, been prolific and successful simply vanished, leaving no descendants to posterity.

I would imagine that the Mesozoic molluscs, Ammonites and Belemnites, were the first fossils most budding palaeontologists added to their collections as children: that certainly was true in my case. I still remember the intense excitement I felt when I was just six years old at finding examples of both on a beach at Charmouth in Dorset! What's more, I still have them, along with the very first fossil I ever found: a Cretaceous echinoid that I discovered among the gravel in the small front garden of my family home.

But again, it is certain that many of the Mesozoic cephalopods were already on a downward spiral before the K-T event. Perhaps it was the metaphorical straw that broke the camel's back for some species . . .

The author displays his first fossils!

Like many children, even at that young age, I was excited but apprehensive about the concept of extinction and read everything I could get my hands on that was even vaguely relevant. It's a fact that virtually every Christmas and birthday present I was given between the ages of seven and twelve was a book about prehistoric life: I still have those, too!

Twelve years later, as a sixth-former studying Zoology, I recall regurgitating the conventional explanation for the demise of the dinosaurs in an 'S' Level exam answer. (At that time 'S' levels were only attempted by students intending to apply for places at Cambridge or Oxford: they weren't the attendance certificates they now seem to have become!)

Among the then-accepted reasons for the Cretaceous extinction were:

The disappearance of the Tethys Sea, a shallow body of water that once existed on remnants of the super-continent of Gondwanaland. Global sea levels dropped by hundreds of metres during the late Mesozoic, and new ocean regions created by continental drifting brought about the draining of the Tethys Sea. This was held to be the reason for the gradual disappearance of the giant marine reptiles, as well as many of the cephalopods.

- Tectonic movements created land-bridges that allowed mass migration of many species of land animals. These brought with them new pathogens for which the resident population had no natural immunity. (A modern example is the speed with which European diseases such as measles and chicken pox decimated the Native American population or the potential threat from tropical plagues such as Ebola which could be transmitted rapidly from one continent to another by air travellers.)

- The appearance of flowering plants, many of which contain powerful toxins, resulted in the extinction of many species of herbivorous dinosaurs through mass poisoning. The main contrary argument to this would be that Mesozoic herbivores, were Darwin's basic principles correct, could surely have evolved digestive mechanisms to deal with this. And, of course, many poisonous modern plants are simply ignored by herbivores.

- Dinosaurs first appeared during the Triassic Period, around 230 million years ago. They were an incredibly successful group that evolved to fill every habitat across the planet. By the end of the Cretaceous (it was believed) many had become over-specialised, occupying very limited ecological niches: this put them at particular risk from climatic and environmental change. (As is the case today with large mammals such as elephants)

- Many dinosaur groups (both sauropod and theropod) evolved into giant forms by the end of their 'reign on Earth'. It was believed that these huge creatures had become so large that even tiny reductions in the amount of available food would have resulted in a population collapse.

The consensus at the time I took my exams was that it was the synergistic result of combinations of some – or even all – of these factors that brought about the Cretaceous-Tertiary extinction.

(I should just say here that the Cretaceous/Tertiary extinction is generally referred to as the K-T extinction [the 'K' is from 'Kalkstein' the German word for limestone] but increasingly is known as the K-Pg extinction, from 'Kalkstein' and 'Palaeogene'.)

Even at the age of seventeen, however, I could see some obvious problems with this idea. For a start: why did some large marine reptiles die out, while others (turtles and crocodiles for example) did not? The same is true of invertebrates: what caused many species to disappear, but *not* squids and nautiluses? And why did some existing groups start to flourish *after* the

Cretaceous extinction? (Corals, mammals and birds, for example.) As for 'over-evolution': many modern species such as Blue Whales, Aardvarks and Giraffes continue to thrive, despite reaching great size or becoming highly specialist feeders – or they would do, were it not for man's influence!

Back in the sixties, a commonly-expressed factor in the extinction of the giant sauropods was that they grew so large that they took most of the day to warm up and become active! This, with hindsight, seems laughable to us today: by the time the core temperature of a Brachiosaurus had reached optimum, its outer layers would literally have been cooked by the Sun's IR radiation! Of course, this (and many other similar notions) was based upon the fallacy that the dinosaurs were essentially reptiles and, like lizards and snakes, being poikilothermic could not maintain the temperature of their blood at a constant level. There is abundant evidence that this was not the case and that, like birds and mammals, the dinosaurs remained active twenty-four hours a day. Furthermore, recently discovered fossils indicate that many – if not all – were covered with insulating feathers or hair. So if the 'Alvarez event' had not occurred, it seems likely that dinosaurs could even have survived the later waves of ice ages.

So, given that it is now generally accepted that birds are, in fact, smallish surviving dinosaurs, and that their much larger Mesozoic relatives were perfectly able to survive even quite dramatic climate changes, the questions that needed to be addressed were 'What killed them off?' and 'Why did birds and mammals survive?'.

Spurred on by the discovery that a *sudden* disappearance of a large percentage of the Earth's plant & animal species had occurred at the close of the Cretaceous Period, palaeontologists started to look for evidence of similar events. Further research quickly revealed that a number of mass extinctions as – if not more – severe had occurred before the 'Age of the Dinosaurs'.

Despite considerable initial resistance to the discoveries of the Alvarez team, palaeontologists began to review the fossil record in a new light and started to concede that the extinctions of many familiar and obviously successful groups may have been misinterpreted. Eventually the search for evidence of the global catastrophes that might have contributed to changes in the pace and direction of evolution began to bear fruit: quite quickly four earlier significant mass extinctions were identified:

The Ordovician-Silurian event. Around 60% of all marine genera (see appendix 4) disappeared around 439 million years ago. It is widely believed that fluctuations in the climate caused the polar caps to expand and then contract, giving rise to a large fluctuation in sea levels.

The Late Devonian Extinction. A similar period of global cooling (evidenced by glacial deposits of this age in Brazil) resulted in the extinction of nearly 60% of marine species, particularly those that are thought to have inhabited warm, shallow seas.

The Permian Extinction. Occurring 251 million years ago, this event was even more severe than the Cretaceous Extinction, accounting for probably **96%** of the animal and plant species on the planet! To put it another way: all life on Earth today has evolved from just 4% of the forms that existed before 'The Great Dying' as this event is often named!

The Triassic extinction. Largely affecting marine species once more, this event resulted in the extinction of around half of all marine genera: undoubtedly some land animals must have been affected also.

More recently, many palaeontologists have begun to question the traditionally-held view that the extinction of many species of megafauna (for example Woolly Rhinoceros and Mammoth) during the **Quaternary Period** was the result of gradual climate change and over-hunting by humans at the end of the last ice age. The discovery of so many 'flash-frozen' Mammoth carcases in Siberia suggests that a far more rapid cooling process may have occurred, as one result of an event that had more far-reaching effects, ultimately bringing about extinctions on every continent.

Once it became fashionable to discuss the possibility that mass extinction events have played an important part in determining the course of evolution, the next challenge was to establish what had caused them. But once again, despite having put the Alvarez team through the wringer for a number of years before having to accept the reality of the K-T event, the reactionary scientific community gratefully turned to comfortable tried and true explanations, as we'll examine in the next chapter.

Even so, this grudging acceptance of catastrophe as a factor in extinction (and hence evolution) had major implications for the traditional theory of Darwinian Evolution – and for those with strongly-held religious beliefs. The idea that a supreme being, having created the teeming multitudes of life on Earth, should then regularly eradicate a large percentage of them does not fit well with orthodoxy! To accept this notion is to accept that from time to time God makes mistakes that need rectifying! This challenge to faith is arguably one of the causes of the flourishing of fundamental religious groups at the end of the twentieth century, many of which deny the entire concept of evolution.

The fact that the fossil record of the Earth provides ample proof that major catastrophes occasionally occur was accepted fairly quickly by most academics: what was either conveniently forgotten or ignored was the fact that a nineteenth century French Scientist, George Cuvier, had reached this conclusion as early as 1815!

Cuvier's work is still largely unknown to the general public, but his researches into geology, palaeontology and classification, as well as his theories on the mechanisms of extinction, laid the foundations of modern thinking.

As is also the case with the equally overlooked British Zoologist, Alfred Wallace, Cuvier's work has been largely eclipsed by that of Darwin. His controversial thoughts on race cannot have helped Cuvier's claim to a place in posterity (he believed that there were three distinct varieties of humans with different mental and physical characteristics) but in this he was merely reflecting the generally held conviction of polygenism.

In my view, a total reassessment of this great *savant's* contribution is well overdue!

Baron Georges Cuvier

Chapter 5

THE SMOKING GUNS OF EXTINCTION

HAVING realised that they had wrongly dismissed the previous work of Cuvier and been beaten to the punch by Walter and Luis Alvarez, those palaeontologists and geologists who didn't go into denial attempted to redeem themselves by identifying the possible causes of global extinction events.

The Alvarez team had reached the conclusion that the K-T extinction had occurred as a result of the massive climate-changing impact of an object from space. The evidence for this was pretty much undeniable:

- High concentrations of the rare metal iridium were found in a thin layer of fine clay deposits that marked the boundary between rock strata of Cretaceous and Tertiary age. These deposits had over one hundred times more Iridium than is typical in Earth rocks. Iridium does, however, occur in meteorites.

- Rocks from the K-T boundary were found to contain spherules or microtektites (formed when small blobs of molten rock cooled and fell back to Earth) as well as shocked quartz grains.

- Many of the K-T clay deposits are dark in colour because of their high carbon content: this is considered to be soot formed from the combustion of most of the Earth's land plants as a result of heat pulses following a major impact.

- Evidence of a mega-tsunami around the Gulf of Mexico at that time provided additional proof of an ocean-impact, as well as suggesting a location for the event.

- The discovery of the buried 180 km Chicxulub Crater on the coast of Yucatán, Mexico was, to most people, conclusive proof that the K-T extinction was precipitated by an extraterrestrial impact.

Micro-tektites from Hell Creek

The scenario for the K-T extinction, as generally agreed today, is that around sixty-five million years ago a large extraterrestrial object (usually identified as an asteroid) smashed into the Earth in the region of what is now the Yucatan Peninsula, Mexico. Sedimentary deposits from around the Caribbean and Brazil suggest that the impact occurred in the ocean, creating mega-tsunamis and projecting pulverised rock dust high into the atmosphere. It has been suggested that sulphuric acid aerosols were simultaneously generated through the sheer bad luck (if you had been a dinosaur!) that the impactor landed in submarine gypsum (calcium sulphate) deposits. These aerosols, together with fine dust, lingered in the upper atmosphere for many months, reducing incident sunlight by a considerable degree and falling as acid rain: this would have stressed damaged ecosystems beyond the point of no return.

Additionally, hot ejecta from the impact would have rained down on the Earth: this (combined with heat pulses from the original impact and the higher partial pressure of atmospheric oxygen at the time) may well have caused the planet's entire plant biomass and most exposed animals to combust. Evidence for this comes from the sooty deposits that define the K-T boundary in many places. Interestingly, the sediments above the K-T layer predominantly contain fern spores, suggesting that these resilient plants were the first to grow in the ashes that covered the land surface.

The effects of the long-lasting aerosol/soot contamination of the atmosphere would have been similar to the apocalyptic aftermath of a nuclear war: a great percentage of the Sun's light and heat would have been reflected back into space or prevented from reaching the Earth's surface: temperatures would have plummeted and photosynthetic organisms would have etiolated and died. Without their essential 'producer' level, many eco-systems on land and, to a lesser extent, in the oceans would have collapsed: first green plant populations would have crashed, followed by herbivorous animals and then the carnivores that preyed upon them. Bear in mind that in any population, the most abundant organisms are the producers, while top carnivores are the least numerous: in a lake there are far more small roach than there are pike: a large predator requires a large reserve of potential prey, given that its victims are generally drawn from the weak, the young and the old.

No doubt, given the amount of dead bodies lying around, medium-sized carrion-eaters would have survived longest among the dinosaurs: but the species with the greatest chance of surviving the 'nuclear winter' would have been those that were small, insulated, mobile and with a fairly non-specific diet. With so many large, rotting carcasses, there were presumably lots of flies, beetles and worms available, while resistant seeds and nuts would have remained edible for years. For these reasons, the post-K-T landscape would fairly rapidly have been repopulated by mammals (which had experienced a previous diversity spurt in the Jurassic) and birds, these effectively being small, agile dinosaurs.

In the seas and large freshwater bodies vertebrates such as turtles, crocodilians and, of course, fish survived, while some previously successful groups did not. Generally speaking, however, the impact of the K-T event on aquatic life was less severe, possibly due to the 'cushioning effect' of water as a medium. Those extinctions that did occur were probably connected to reduced oxygen levels.

It is well-known that large reptiles (and we may include snakes and lizards in this) have the ability to survive long periods without food: they may hibernate, become torpid or reduce their metabolic rate. Furthermore, many crocodilians seem actually to require that their prey be in a semi-decomposed state before eating it. In this way, some archaic groups may have survived until vestigial populations of algae and phytoplankton re-established aquatic ecosystems.

Sooty K-T material from Alberta

However, while many palaeontologists and astronomers were eventually persuaded that the K-T extinction was (at least in part) the product of a massive impact by an extraterrestrial object, others expressed doubts or suggested other mechanisms for catastrophe.

Call me a cynic, but it has always occurred to me that iconoclasty is a great way to achieve media attention, attract funding, or get a book published! If you were to promote a theory, for example, that modern humans are descended from hominids that spontaneously and simultaneously evolved in several regions of the Earth, rather than just eastern Africa (and could provide evidence that apparently supported your contention) you could expect at least to attract the attention of the more right-wing media.

So it has proven to be with the idea that many (if not all) major extinction events were the product of extraterrestrial impacts. No sooner had the Alvarez team achieved general acceptance that Chicxulub was the site of the impact that caused the K-T extinction than researchers began to search for – and find – apparent flaws in the proposition or alternative mechanisms:

- Other large craters were discovered that were not obviously associated with an extinction episode. This brought into question the idea that a single impact could cause devastation on a global scale.

- Extinction events were identified that were not dated to the ages of known large craters

- At around the same time as several mass extinctions took place, massive flood basalt eruptions occurred: these were implicated as the cause of devastating 'Nuclear Winters' that either caused or contributed to global catastrophe.

- The powerful anthropogenic climate change lobby have been keen to recruit the K-T extinction to their cause, as evidence of how increased carbon dioxide emissions can and have resulted in the collapse of global ecosystems. It has been suggested that the eruption of super volcanoes and flood basalts periodically result in the release of climate-changing volumes of carbon dioxide.

- Some palaeontologists continue to contend that the K-T event was not as devastating as generally agreed, and that further excavation will eventually produce dinosaur fossils from the Palaeogene. (The modern identification of birds as surviving dinosaurs helps foster this suggestion, as does the widespread belief in semi-mythological dinosaur-type creatures such as M'boko M'bele and the Loch Ness Monster.)

Let's consider these (like some curious palaeontological beauty pageant!) in reverse order!

As Sherlock Holmes said in *'A study in Scarlet'*:

"It is a capital mistake to theorize before you have all the evidence. It biases the judgment."

It is surely faulty logic to base an argument **against** the idea of global catastrophes as a mechanism for abrupt global extinction upon what evidence **might** be unearthed in the future! At the time of writing, no authentic fossils of dinosaurs (other than birds) have been discovered above the K-T boundary layer. And, despite diligent searching (and dwindling equatorial forests in Africa) no-one has yet proven that a single monster legend from folklore is based on a surviving dinosaur.

Whether you believe that human beings are having a measurable effect on global climate or not (and many reputable scientists do not!) it is surely somewhat bizarre to support a cause for the Mesozoic extinctions purely because it appears to highlight the danger of increased carbon dioxide emissions from cars and factories! What is not widely acknowledged is that atmospheric CO_2 levels rose from around 420 ppmv (parts per million by volume) in the Triassic to a peak value of 1,130 ppmv by the middle of the Cretaceous. The result of this was a proliferation of foraminifera and molluscs that utilised carbon dioxide to create their shells. The vast amount of calcium carbonate that was then deposited when the organisms died produced the huge quantities of chalk, limestone and dolomite rocks we find today. It's hard to square this gradual increase in atmospheric CO_2 with the sudden emissions that would result from volcanic or flood basalt eruptions and equally unrealistic, in my view, to contend that they were the cause of the K-T extinction.

Another interesting problem is that of providing a credible reason for the coincidental eruptions of flood basalts at the same time as several well-evidenced impact events.

The devastating Permian extinction occurred at around the same time as the eruption of the Siberian Traps and the published formation dates of vast impact craters in Wilkes Land, Antarctica and Bedout High, north-western Australia. Similarly, the date of the eruption of the Deccan Traps in India roughly coincides with that of the impact of large extraterrestrial crater-forming bodies at Chicxulub, Silver Pit (in the North Sea), Shiva (India) and Boltysh (Ukraine)

Although it must be acknowledged that the ages of some of these craters remains contentious at the time of writing, it wouldn't be too surprising for large multiple impacts to occur simultaneously: this was demonstrated by the fragmentation of comet Shoemaker-Levy 9 and the subsequent collision of the remnants with Jupiter in 1994. Any one of these impacts would have generated the kind of environmental catastrophe we are attempting to understand.

But here's a thought: what if a *really big* chunk of our hypothetical multiple K-T impactor not only produced the craters identified above, *but was also responsible for the eruption of the Deccan Flood Basalts?* How might this have occurred?

The Earth's crust, as we have discussed earlier, exists in two forms: oceanic crust and continental crust. Some regions of continental crust are geologically active and tend to be around 100km in thickness, while other, more stable regions known as cratonic crust, can be double that figure. Oceanic crust is denser but has a maximum thickness of just 10km.

The Deccan Traps are a sub-aerial volcanic outflow that covers much of Maharashtra in west-central India. It is generally agreed that a hot mantle-plume eroded a channel to the surface around 66 million years ago, releasing vast amounts of potentially climate-changing gases and covering 1.5 million km² of the surface with basalt. But could such a hypothetical plume manage to burst through a region of continental crust? Or was the Earth's crust at this point weakened by a massive impact – possibly even a fragment of the same body responsible for the Chicxulub crater? It is surely not too fanciful to suggest that a huge impactor striking the ground at the end of the Cretaceous Period compromised the crust to the extent that magma was able to travel upwards creating the traps.

If such a devastating event happened in India, might not similar impacts be responsible for other flood basalt eruptions, including that which occurred at the close of the Permian Period, forming the Siberian Traps? It is the case that the Wilkes Land Crater would have been directly opposite the region of the Traps on the Earth's surface when both were formed: could there be a link? It seems an incredible coincidence if not!

The fact that large craters have been identified that are not associated with extinction

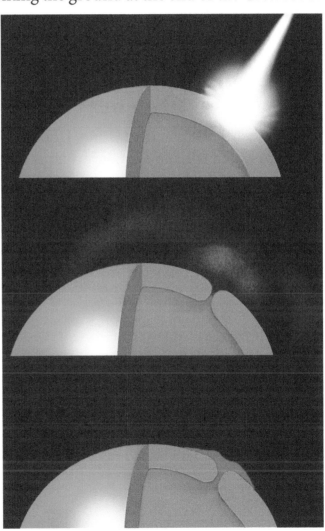

Possible formation of flood basalts

events may be explained quite simply: firstly, many of these craters are simply not big enough to have affected ecosystems other than those close to the impact site. More importantly, it appears that even a massive impact may not on its own be sufficient to cause truly global devastation. As mentioned earlier, it is widely believed that the effects of the Chicxulub event were exacerbated by the injection of vast amounts of sulphuric acid into the upper atmosphere: not only would this have contributed to rapid and sustained global cooling, but ecosystems that survived the initial threat would have been compromised by prolonged acid rainfall. And, as we have noted above, impact events seem to have their greatest effect upon ecosystems that are already under other pressures.

The fact that some extinctions (though not the major events!) seem not to be associated with any known crater is similarly easy to account for. Firstly, a crater may exist, but just hasn't been identified yet: it may be buried under younger strata, be hidden beneath the polar ice caps or it may be underwater (large areas of continental shelf were flooded by rising sea level following the last sequence of ice ages) A second possibility, which we will explore in chapter eight, is that an incoming 'extinction object' exploded before it reached the Earth's surface *without leaving a crater!*

Chapter 6

WHAT MADE ALL THOSE BIG HOLES?

AS early as 1903 mining engineer Daniel Barringer concluded that the 1.2km wide Canyon Diablo Crater in Arizona had been formed by the impact of an iron-rich extraterrestrial object. As is widely known, Barringer established a company to exploit the vast amounts of iron minerals he was convinced lay in and around the crater: in this he was less than successful!

However, it was not until 1960 that astronomer Eugene Shoemaker presented evidence to prove to virtually everybody's satisfaction that the Barringer Crater (as it was known by then) was meteoric rather than volcanic in origin. Apart from the appearance of the crater (one of the most impressive then known) Shoemaker's most compelling evidence was provided by the presence at the site of the shocked quartz minerals stishovite and coesite. Neither of these are common on the Earth's surface (but may be super-abundant in the mantle) nor are they associated with vulcanism: both are now considered to be indicators of the immense energies released by a massive impact event.

The impactor that produced the Barringer Crater was a nickel-iron mass of around three hundred thousand tonnes, the greater part of which vaporised on impact: small pieces of the meteorite are a familiar item in museums and collections.

Once the metaphorical floodgates had been opened, geologists and meteoricists hunted for – and found – more and more impact craters, ranging from vast astroblemes like the Sudbury Ring and Manicouagan Crater in Canada to smaller, better defined structures such as Wolfe Creek in Australia and Gebel Kamil in Egypt. The latter was discovered in 2008 by Italian researcher, Vincenzo de Michele while exploring the Sahara Desert on *Google Earth*! This recent search tool adds new impact sites on an almost monthly basis, so that the number of known or suspected impact craters

identified has reached nearly 200 at the time of writing, with probability calculations suggesting a further 340+ remain to be discovered.

As a meteorite lecturer and dealer, I am always interested in new craters, not least because they may be a source of new material for my inventory! Gebel Kamil, for example, has yielded around 800 kg of attractive ataxites (iron meteorites that do not possess the characteristic crystalline Widmanstätten pattern that the majority display).

A year or two ago I was asked to write a piece about meteorites and impact craters for a mineralogical magazine. As I reviewed the data, it occurred to me that only a small proportion of meteorites available to the trade or collectors actually originated from known impact sites: what's more, none have been found at any of the very largest craters. Assuming that iron meteorites are more likely to pass through the atmosphere unscathed, it is curious that the majority of well-known falls or finds are located in shallow strewn fields in which the individual craters are usually just a few metres across. (The passage through the atmosphere and impact of some meteorites is observed at the time it occurred: retrieved material is referred to as being from a *fall*. Others are chanced upon by meteorite hunters and are known as *finds*.)

A good example of a large strewn field with no large associated crater is the Campo del Cielo find. The original huge iron mass fell approximately 6000 years ago in Argentina, about 500 miles NNW of Buenos Aries. It seems possible that the object either exploded high in the atmosphere or actually arrived as a shower of fragments, since the remnants are scattered over a wide area.

In 1576 a Spanish governor of the region was told of the iron masses by local people, and sent an expedition to search for them. The leader, Captain de Miraval, discovered the strewn field at a location known by the indigenous population as the *Campo del Cielo* ('field of the sky or heaven'.) Presumably this name was given to the region by local people who had witnessed the fall of the object.

Despite the export of many hundreds of small examples of this popular meteorite, thousands more remain, including some truly huge individuals. Intriguingly, the only craters at the site are shallow depressions a few metres across.

This is also the case with the majority of the large, better-known strewn fields: no sizable crater is associated with Sikhote Alin, NWA 869, Gibeon or Muonionalusta. Bearing in mind that the object that created the Gebel Kamil crater is estimated to have been less than two metres in diameter, it is hard to explain *the lack of any meteoritic material associated with the really large craters.*

To put it another way, the currently received wisdom about the large craters generally acknowledged to be associated with extinction events seems somewhat flawed, to say the least:

- Some (if not most) mass extinctions were caused by the impact of huge solid bodies: large meteoroids or asteroids.

- These impacts created immense craters, releasing vast amounts of energy and throwing climate-changing quantities of pulverised rock into the atmosphere.

- The impactors responsible for these extinction events would have been kilometric in size.

- No remains of the impacting bodies responsible have been found.

- Many smaller craters (less than 5km in diameter) have produced quantities of meteoric iron.

- Large stony or nickel-iron meteorites have frequently fallen to Earth without leaving any crater at all.

How can this be? In fact, there are several reasons why meteoroids – even large ones – can *almost* float down to the Earth's surface! Suppose a meteoroid 'creeps up' behind the Earth in its orbit around the Sun: the closing speed might be just a few hundred kilometres per hour. This would naturally increase as the Earth's gravity accelerated the object inwards, but even so the kinetic energy of a 'sneak attack' would be considerably less than a head on collision, where the closing speed could easily be two hundred thousand kilometres per hour!

Amazingly, even the time of a meteoroid's arrival has an effect on its kinetic energy! Since the Earth spins on its axis from west to east at around 1600kph, a meteoroid entering the atmosphere at dawn (that is, on the eastern limb of the Earth) would have its entry velocity reduced by that amount!

It should be explained at this point what is generally meant by the word 'meteoroid'. Any small stony or metallic object travelling in an orbit around the Sun is a **meteoroid**: the upper size limit would perhaps be ten metres. If such an object entered the Earth's atmosphere it would become extremely hot due to friction, leaving the visible trail we call a **meteor**. If the original object were large enough (the size of a football) it might reach the Earth's surface: it would then be described as a **meteorite**.

Some ancient meteoroids accreted directly from material in the solar disc and would therefore be among the oldest objects in the Solar System: others seem to be debris from collisions between planetissimals, planets and asteroids. (This group includes planetary achondrites such as the SNC group from Mars, Lunaites that originated on the Moon and HEDs, many of which seem to have resulted from a massive impact on the asteroid 4-Vesta.) Other meteoroids, as we shall explore in more detail later, are the remnants of extinct comets: these are generally responsible for the periodic meteor showers.

So what, then, are asteroids?

Whenever the media search for a name to tag on to any large object that has collided with the Earth in the past or may do so in the future, the one they invariably arrive at is 'asteroid'. For example: the smallish fragment of rock that arrived on Earth in 2008 at Almahatta Sitta in the Nubian Desert of Sudan was immediately labelled as an asteroid in the newspapers and on television. The same is true of the 10,000 tonne object that exploded above the Russian city of Chelyabinsk in February 2013. I find it both amusing and frustrating that a well-known online encyclopaedia defines asteroids as being objects larger than 10 metres yet publishes a map of the world showing the arrival positions of *'asteroids between one and twenty metres that disintegrated in the Earth's atmosphere'.* This seems to me to be a serious case of bet hedging!

In fact, to astronomers, the term asteroid has a quite specific meaning: these objects are essentially minor planets that are predominantly found in fixed regions of the Solar System. The most populous of these are:

Main Belt asteroids. These orbit the Sun in a region between Mars and Jupiter. It is estimated that there are up to two million main belt asteroids larger than 1km in diameter, with many much smaller objects. It is widely believed that the Main Belt is a region where planetary accretion was prevented by gravitational interference by Mars and Jupiter. By and large these objects are in stable, non-eccentric orbits that would never intersect that of the Earth.

Trojan asteroids co-inhabit the orbit of a larger planet but, occupying two stable regions of the orbit (known as Lagrangian Points) will never collide with it. It has been suggested that Jupier's Trojans may be as numerous as those of the Main Belt

Near-Earth asteroids are generally somewhat smaller (about 1,000 NEAs over 1km in diameter are known at the time of writing.) Of the 12,000 NEAs so far identified, a few are designated as 'Earth-crossing Asteroids': that is, objects over 10 metres in diameter whose orbits *may* intersect that of the Earth.

An interesting point to keep in mind is that part of the definition of a planet is that it should have 'hoovered up' all the material in its region of space. This was one of the main reasons for relegating Pluto from the list of nine planets we all learned in school: its orbit is littered with other objects, some of which are themselves getting on for being planet-sized. So why do these 'Earth-crossers' exist? In my opinion they are either in orbits that have never – *will never* – bring them into collision with our planet, or they are objects whose orbits have been perturbed at some time since planetary accretion was completed.

The lecturer and author David Icke has identified the political principle of

PROBLEM : REACTION : SOLUTION

To me the continual barrage of media suggestions that a major extinction-event sized asteroid impact could occur at any time without warning is a classic example of this philosophy. The likelihood of such a thing happening has been hyped sufficiently to create a clamour for further space research, better radio and optical telescopes and for governments to come up with a plan. In the United States, this has led to the enfranchisement of organisations such as the Federal Emergency Management Agency, while in the United Kingdom the Department for Communities and Local Government would probably co-ordinate the response to a major extraterrestrial impact event.

In fact (Bruce Willis notwithstanding) an incoming asteroid the size of the object responsible for the Chicxulub event could not possibly be deflected or fragmented using current technology. Furthermore, one would have to question the wisdom of shattering an asteroid into millions of car-sized chunks, each of which could potentially make a Gebel Kamil-sized crater . . .

POTENTIAL IMPACT REPERCUSSIONS

15m meteoroid : 1 Megaton (MT)
Serious local consequences, though atmosphere provides partial shield. Hydrogen-bomb scale, but without the radioactivity.

1km asteroid: 250,000 MT
No atmospheric shield. Hemispheric-scale effects. At threshold for global effects. Significant fraction of all humans killed.

10km asteroid: 250 million MT
Global effects. Ejected, vaporized rock and water fill atmosphere resulting in global winter and major extinction of lifeforms, including virtually all humans.

My personal belief is that the risk of an asteroidal impact on Earth is comparatively low and that, since the majority of asteroids are less than 20 metres in diameter, damage from most such impacts would be restricted to a relatively small region.

The logic, to me, is inescapable: the biggest craters on Earth (and, by inference, elsewhere in the Solar System) *were not the result of meteorite or asteroid impacts!*

Let's see if we can identify the actual culprits!

The Bayeux Tapestry

Chapter 7

HARBINGERS OF DOOM!

MOST people are probably aware of the Bayeux Tapestry, an astonishing piece of eleventh-century embroidery made, it is generally conceded, on the orders of Bishop Odo, brother of King William 1 (William the Conqueror.) Its fifty or so scenes depict events leading up to the Battle of Hastings and the Norman Conquest, possibly including the death of the Saxon King, Harold 11.

Of more particular interest to our discussions is the panel that depicts the King and his subjects cowering at the sight of a great comet in the skies over Europe. This object has long been identified as Halley's Comet, which (until its disappointing apparition in 1986) returned every 76 years to put on a dazzling display in the night sky. It is known that the comet was included in the embroidery because, during the run up to the Battle of Hastings, it was taken to be a portent that Harold's days as King of England were numbered.

In fact, comets have long been considered as signs of regime change or the death of a ruler: in Shakespeare's play *Julius Caesar* Calpurnia (Caesar's wife) states:

"When beggars die there are no comets in the sky. The heavens only announce the deaths of princes"

However, it has been demonstrated that comets were also interpreted as the passing into heaven of the souls of divinities like Caesar.

The ***Caesaris astrum*** of Julius Caesar was what is generally known as a Great Comet: it possessed a tail that stretched from the horizon to the zenith and was visible in broad daylight. To my huge regret, since my birth in 1951 there hasn't been a comet even approaching this in spectacle, although several have been promised!

Comets (the word arrives by a circuitous route from the Greek phrase for 'wearing long hair') have undoubtedly excited attention and discussion from the earliest days of civilisation: it has even been suggested that the Star of Bethlehem may have been an apparition of Halley's Comet.

A couple of hundred years ago astronomers began to recognise the true nature of comets as comparatively small objects that grow in apparent size and brightness as they approach the inner Solar System, often developing a fuzzy 'head' (coma) and a tail that faces away from the Sun. Some of the most impressive comets may produce multiple tails: the Great Comet of 1861 had at least six! Calculations by astronomers such as Edmund Halley determined that some comets are periodic, with regular and predictable apparitions: others enter the inner Solar System on parabolic paths, swing around the Sun and return to the depths of space never to return. It was also observed that a link existed between some short-period comets and annual meteor showers. (For example, the Perseid Shower is linked to the comet Swift-Tuttle, while the Eta-Aquarids and Orionids are both associated with our old friend Comet Halley!)

These two observations (and some logical thought) suggested that comets probably consist of a volatile core with an outermost layer of stony material:

this model has since been confirmed during close fly-pasts and landings by spacecraft such as Giotto and, more recently, Rosetta / Philae. The core of a comet is currently believed to have a small rocky centre, surrounded by layers of frozen water and lesser amounts of frozen ammonia, carbon dioxide, carbon monoxide and methane. Coating this is a regolith of dark, organic material and stony debris that has been collected during the comet's journey through space.

As a comet approaches the Sun, the icy material in its nucleus begins to sublimate, producing the characteristic coma. The solar wind may push dust particles away from the nucleus forming a ***dust tail***, while gases ionized by solar radiation may produce a second ***ion tail***. Eventually, after a number of encounters with the Sun, a comet may lose most of its volatiles, leaving a cloud of rock and dust that can no longer generate a coma or tail: this seems to have happened to Halley and, more recently, to the much-anticipated Comet ISON in 2015.

Given the current model of comets as 'dirty snowballs', it is interesting to conjecture how they originally became associated with death and destruction.

A novel and contentious suggestion was proposed, by Sir Fred Hoyle and Dr Chandra Wickramasinghe among others, as part of the theory of panspermia. This postulates that unexpected pandemics (such as the 'flu epidemic of 1918 and various plagues throughout history) are caused by viruses that are carried around the cosmos on comets. This idea was firmly believed by Chinese philosophers as far back as 1500 B.C. The "Mawangdui Silk" listed 29 different cometary forms and the various infections, disasters and social upheavals associated with them. In 648 AD, the imperial astronomer Li Ch'un Feng wrote:

"Comets are vile stars. Every time they appear in the south, they wipe out the old and establish the new. Fish grow sick, crops fail, Emperors and

common people die and men go to war. The people hate life and don't even want to speak of it."

In the search for the possible origins of life on Earth (and other planets) aperiodic comets may yet be discovered to be a major vector for the transport of nucleic acids around the cosmos: we are only just beginning to understand the significance these objects may have had – and may continue to have – on mankind and the development of life on Earth.

But plague-carriers or not: might comets represent the greatest threat to mankind's continued existence?

Chapter 8

THE THREAT FROM ABOVE:
MANKIND'S NEMESIS?

IN the previous chapters we moved towards the conclusion that meteoroids and asteroids are unlikely to have been the objects that excavated the large number of craters that pock-mark the surface of every stony planet and their satellites. Furthermore, all the evidence suggests that they were not the impactors responsible for a number of global extinction events: in the case of meteoroids, these by definition are not large enough to create devastation on a global scale, while the chances of a collision with an Earth-crossing asteroid any time after the Late Heavy Bombardment would seem to be diminishingly tiny: comets, however, are much more likely candidates!

The majority of comets within the Solar System inhabit two well-defined locations: the Kuiper Belt and the Oort Cloud. The former is a disc-shaped region beyond the orbit of Neptune that contains numerous small icy planetoids as well as billions of comets. The Kuiper Belt is considered to be the origin of short period comets which take less than 200 years to complete an orbit of the Sun, while many long period comets are derived from the Oort Cloud. This is much more distant: around 100.000 Astronomical Units.

(One AU is the distance from the Earth to the Sun: approximately 150 million kilometres.)

It is assumed that comets and other bodies within these two regions condensed from the Solar Nebula at the same time as the planets: around four and a half billion years ago. However, it is distinctly possible that the cometary populations of both might be augmented by ice masses entering the Solar System from deep space. Water molecules are two-thirds hydrogen, the most abundant of all elements: oxygen has an atomic number of eight and was also present in the early stages of the Universe. From this we can

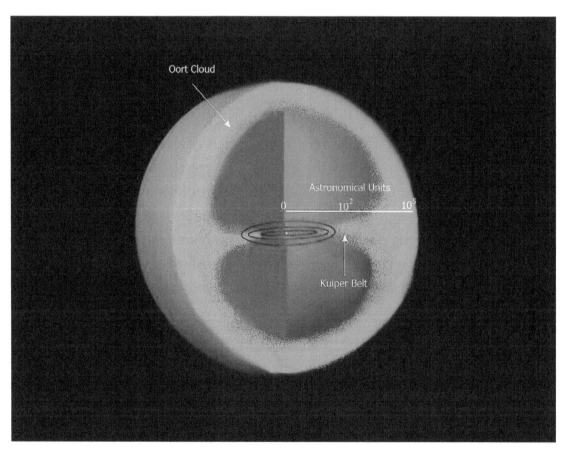

Location of the Oort Cloud and Kuiper Belt

infer that water in one state or another is extremely common, not just in planetary systems, but in interstellar space. It seems at least feasible that some of the erratic comets that arrive unexpectedly on parabolic paths may enter our Solar System from the void between the stars. This, of course, makes it impossible to predict their appearance!

At the time of writing, no-one has come up with a wholly credible explanation of why a comet should suddenly leave the Oort Cloud and travel towards the Sun. One possibility that has recruited a lot of support within the astronomical community is that occupants of the Oort Cloud are occasionally perturbed by a passing brown dwarf star. To me this seems a bit of a fudge! Although Brown Dwarfs (small stars with too little mass to become self-luminous) are quite likely to be among the most abundant large bodies in the Universe, there is absolutely no evidence of a population of them near enough to affect the Oort Cloud. This is, perhaps, not surprising: being the size of a very large gas-giant planet and not emitting any light, it would be very hard to detect a Brown Dwarf directly. As far as I am aware, nobody has yet found one lurking on the outskirts of the Solar System (the nearest being Luhman 16 at a distance of six and a half light years). It would therefore seem a little too far-fetched to suggest such a body as a candidate for the dramatically named 'Nemesis Object' that is widely considered responsible for occasionally tipping comets into the inner Solar System.

Let's pull all this together and try to reach a conclusion that satisfies the principles of Occam's Razor: the simplest solution is usually the correct one!

- The Earth and every other rocky world in the Solar System bear the marks of innumerable high-speed impacts by several types and sizes of object.

- Throughout the history of the Earth, particularly devastating impacts have been responsible for a number of global extinction events.

- Although the bodies responsible for these catastrophic impacts have long been identified by the media and Science alike as meteoroids and asteroids, this does not seem a credible thesis: examination of all large impact sites has failed to produce any trace of a stony or nickel-iron impactor.

- While unusually high levels of Iridium are often found in rock layers associated with large impact sites, it is known that the regolith of a large comet is rich in this rare element.

- There are countless cometary nuclei orbiting the Sun within the Oort Cloud and Kuiper belt: the possibility exists that additional comets arrive continuously from deep space.

The kinetic energy of a ball of ice a kilometre across travelling at a hundred thousand kilometres an hour (around 2,777 metres/second) is unimaginably vast: the formula for calculating it is:

$$K = \tfrac{1}{2}mv^2$$

(K = kinetic energy, m = the mass of an object, v = its velocity)

A ball with a radius of 500m has a volume of 523,809,524 cubic metres: if it is made of ice, its mass would be 502,857,143 tonnes. Travelling at the velocity above, its approximate kinetic energy would therefore be:

½ x 502,857,143 x 2777 x 2777 joules

. . . which works out at **1,939,499,843,500,000 joules**. That's a lot!

A one megaton nuclear device produces **4,184,000,000,000** joules of energy, so that a comet just one kilometre across hitting the Earth would release energy equivalent to a **200 megaton** nuclear explosion! That's about the same as a thousand bombs the size of that which was detonated over Hiroshima in 1945. (Bear in mind that there are trillions of comets in the Solar System that are *tens* of kilometres across!)

Without doubt, astronomer Fred Whipple's description of comets as 'dirty snowballs' has lulled people into a false sense of security: even a small comet striking the Earth would be entirely capable of producing a mass extinction of the magnitude of the K-T event, while a medium or large cometary impact has the potential to eradicate permanently all plant and animal life.

Recent robotic investigation of comets has provided a much better understanding of their structure and chemistry: using this information, is it possible to identify any major impacts events as cometary?

One of the best-known and most discussed astronomical events of the twentieth century was the huge explosion that devastated a large area of forest near the Tunguska River, in what is now Krasnoyarsk Krai, Russia.

Early in the morning of June 30th, 1908 a large fireball was seen moving across the sky, followed by a brilliant flash of light. Shortly afterwards a loud explosion was heard, accompanied by a shock wave that was sufficiently strong to damage buildings and knock people off their feet even many kilometres away.

It wasn't until 1927 that an expedition was mounted to investigate the explosion (the delay largely being a result of Russia's participation in the First World War and the subsequent upheavals of the Communist Revolution of 1917)

When members of the expedition (led by geologist Leonid Kulik) arrived at the epicentre of the explosion they were greeted by an extraordinary sight: 2,150 square miles of forest had either been burnt or flattened.

To Kulik's surprise, however, there was no sign of a crater at 'ground zero', nor were any meteorite fragments found.

A television documentary broadcast in 2014 followed the adventures of a group of international scientists as they investigated the Tunguska site: each had a pet theory:

- An asteroid or meteoric impact

- An encounter with a chunk of anti-matter

- The explosive release of trapped gases from the Earth's mantle

- Aliens sacrificing their spacecraft to save humanity (No: really!)

One might reflect on the strange fact that these highly-qualified science professionals held such a diverse range of theories, each of which attempted to account for a single event! The explanation for this is surprisingly straight forward: over a century later and following numerous expeditions to Tunguska, there is absolutely no concrete evidence at the site to indicate what caused the explosion!

Apart from the surviving toppled trees, the only tangible 'evidence' presented in the programme was eventually found by a geophysicist. He located a small outcrop of sandstone, samples from which appeared to contain shocked quartz grains. Bizarrely, he claimed these to be proof that an explosive release of mantle gases from beneath the Siberian Traps basalt layer had been responsible for the devastation! As was mentioned earler, shocked quartz is frequently encountered at the site of an extraterrestrial impact: in fact (along with shatter cones) such shocked grains are considered to be the 'smoking gun' of an impact crater: it is axiomatic that they are never associated with volcanism.

Little need be said about the anti-matter and UFO 'theories': there is, as yet, no proof that anti-matter actually exists, except as a theoretically necessary extra mass to provide the gravity needed to hold the Universe together! And sadly (despite the attractive idea of altruistic aliens giving their lives to save some herds of reindeer!) no fragments of exotic or extraterrestrial technology have been located in the region of the explosion.

To state what must, by now, seem an obvious conclusion: the Tunguska event *must* have been the result of a comet exploding high in the atmosphere. Only in this way can we account for all the observed phenomena: the vast release of energy high in the atmosphere, the absence of an impact crater and the failure of numerous search teams to recover *any* meteoric material.

If further support for this explanation is required, it arrived, I would contend, over the Russian city of Chelyabinsk in February 2013.

In a remarkably familiar reprise of the Tunguska event (captured on dashboard cameras for the whole World to marvel at) an object entered the atmosphere over Siberia, leaving a dense white cloud before fragmenting and exploding violently. Windows were shattered, people were injured by flying glass and shock waves travelled hundreds of kilometres. Once again, nothing large enough to cause a crater reached the Earth's surface, although on this occasion some solid material *was* recovered.

As a meteorite dealer, as soon as images of the fireball appeared on the TV and internet, I was keen to try and obtain some fragments of the 10,000 tonne object being described by media 'experts'. At the first opportunity, I e-mailed a couple of contacts in the region: their responses both surprised and intrigued me. Yes, they informed me, stony material *was* being recovered, but it was chiefly tiny wholestones. The images of these that were attached showed pea-sized objects with glossy black fusion crusts. Some had been fractured by the fall (or by the children who had them in order to maximise the quantity they had to sell!) and these displayed a beautiful dove-grey interior. The largest examples I was offered had masses around 200 grams, while the majority were between 5g and 10g.

Straight away this seemed unusual: normally, when a large stony meteorite explodes during its passage through the atmosphere (as happened in 1969 at Barwell, Leicestershire) the recovered fragments are irregularly-shaped and display just small patches of fusion crust.

Soon after the explosion the press corps descended upon Chelyabinsk and were shown an eight-metre hole in the frozen surface of Lake Cherbarkul. It was confidently predicted that the main mass of the impactor would be found on the lake bed beneath: it wasn't!

To me there could be only one explanation: the hole had been made by a chunk of ice that had quickly melted on the lake bed. I was contacted by several journalists anxious for a sound-bite and gave my opinion that the Chelyabinsk object had been a small comet. My reasoning was quite straight-forward:

Chelyabinsk meteorites

- The white trail left by the object as it passed through the atmosphere seemed most likely to be water vapour, more consistent with the ablation of an icy comet than a stony meteoroid.

- The thousands of small, complete meteorites that were recovered seemed unlikely to have originated in the disintegration of a single large object: to me it seemed more probable that they had been spherical chondrites embedded in the regolith of a comet.

- Of the estimated 10,000 tonne mass of the original object, only a few dozen kilos of meteorites were recovered, despite diligent searching by hundreds of people.

To my frustration (but not surprise!) my comments were not published, perhaps because they didn't conform to the 'official' view that the exploding object had been an asteroid fragment.

In the following months I was invited to write articles and give a number of lectures about the Chelyabinsk event: naturally I put forward my opinions about the nature of the object responsible. Then, as mentioned earlier in the book, in 2014 I had an opportunity to discuss the matter with Rusty Schweickart, former Chairman of the B612 Foundation. To my surprise his response, like that of many of the people who had attended my lectures, was polite but completely negative: he was adamant that the greatest danger of a catastrophic impact on the Earth came from an asteroid and would not concede that comets had ever played any part in global extinction events. I couldn't help but reflect that there might have been a political agenda in play somewhere!

A final example (and there are many more!) of an anomalous site that is considered by some researchers to be cometary in origin is the Gilf el Kebir, in the Egypt / Libya border region of the Sahara Desert. Whether a large circular astrobleme can be made out on images of the plateau taken from space is, to some extent, a matter of faith! What is certain is that beautiful, variously-coloured natural glasses have been collected from the Gilf since ancient times:

Famously, in the centre of a pectoral jewel found in the tomb of Tutankhamen is a golden Scarab made from Libyan (or sometimes Egyptian)

Desert glass and tektites

Desert Glass. This is taken as evidence that the material was considered something special by the ancient Egyptians, possibly associated with the Sun disc that the scarab-god Kheper pushed across the sky in their mythology.

Many authorities agree that Libyan Desert Glass (LDG) is a glassy impactite formed when a vast aerial explosion fused local sand and limestone around twenty-nine million years ago. On analysis 98% of LDG is found to be silica, the residue consisting of carbonate minerals, spherules of cristobalite (a high-temperature form of silica) and, rarely, dark, brownish streaks. These have been found to contain many of the same minerals as chondritic meteorites, suggesting the material has an extraterrestrial origin.

Meteoricists usually consider Libyan Glass to be one of a number of glassy impactites known as *tektites*. Many tektites show signs of high-speed ablation, and it has sometimes been suggested that they have an extraterrestrial – possibly even lunar – origin. (See papers by geologists John O'Keefe and Hal Povenmire) Other tektites have stratified structures and inclusions like those of LDG and are considered more likely to have been formed as thermally-fused slabs below an atmospheric explosion.

Perhaps significantly, neither the aeroform tektites nor the stratified Muong Nong types have been conclusively associated with a crater! The Indo-Chinese tektite field is absolutely vast and has produced literally millions of Tektites, yet no crater has yet been found that can definitely be identified as its source. Here again we have evidence of a dramatic event high in the atmosphere that generated extreme temperatures at ground level.

So, having reviewed the evidence currently available, can we identify the past and future cause of global extinction events? I believe we can ...

Beyond the furthest reaches of the Solar System, at a distance between 5,000 to 100,000 AU, lies an encompassing shell of trillions of cometary nuclei: the Oort Cloud. Since their formation from the solar nebula or capture by the Sun, these objects have collected a regolith of carbonaceous material and dust particles.

For reasons not yet fully understood (but possibly associated with the gravitational influence of passing dwarf stars) the orbits of these bodies can be perturbed, causing them to tumble inwards towards the Sun. Should this occur, by the time an Oort Cloud object reached the inner Solar System it would be travelling at a velocity of one hundred thousand kilometres an hour and will have acquired an outer shell of rocky material.

Encountering the electromagnetic radiation and streams of particles from the Sun, the icy nucleus may develop a coma and tail, becoming visible from the Earth as a typical comet. In common with the great majority of these objects, it may swing around the Sun and pass harmlessly back into deep space. Possibly, however, it will be captured and enter an elliptical orbit, becoming a periodic comet that makes regular returns to the inner Solar System. Eventually it will lose most of its mass through sublimation, leaving a cloud of stony debris that might give rise to a new meteor shower, should its orbit cross that of the Earth.

Just occasionally (perhaps once every fifty million years) the downward plunge of a large comet into the inner Solar System might bring it on to a collision course with our planet. As it enters the Earth's atmosphere, the temperature of the comet's outer layers rises rapidly to thousands of degrees Celsius, while that of its inner core remains close to absolute zero. A combination of this asymmetric heating and the physical stresses of sudden deceleration will cause the comet to explode violently. If it is a kilometre or more in diameter, the kinetic energy of its great mass and velocity would be converted into a devastating release of heat, light and pressure that would throw trillions of tonnes of pulverised rock high into the atmosphere. The heat pulse and shock wave of the explosion would travel around the globe, instantly vaporising lakes and pools and incinerating all combustible material. This is the familiar scenario of a global extinction event ...

At the time of writing, several serious-minded researchers have begun to link cometary impacts with another potentially devastating possibility: the giant tsunami.

(Strictly speaking, a tsunami is a destructive wave caused by a sub-sea earthquake, such as that which occurred in the Indian Ocean at Christmas, 2004. In recent years the Japanese word has replaced the previously used term 'tidal wave' to differentiate water displacements caused by seismic events from those produced by storm surges or unusually high tides.)

The history, mythology and archaeology of mankind suggests that several times in the last ten thousand or so years whole civilisations have been destroyed by vast incursions by the sea. The biblical story of Noah, the Babylonian Epic of Gilgamesh and the Indian legend of Manu and Matsya are just the best known of dozens of similar accounts from every continent. It seems likely that many – if not all – of these are based upon actual events: archaeologists have suggested that the cataclysmic eruption of Thera was the mechanism for the destruction of the Minoan Civilisation on the island of Crete and that the inundation of the Black Sea basin around 5,600 BC by glacial melt-water may also have generated enduring flood mythologies.

Increasingly, however, studies of marine sediments both underwater and on dry land have indicated that large impacts in the world's oceans have been responsible for even more devastating tsunamis. The most credible evidence for this comes from the large deposits of sediment (known as 'chevrons') that have been identified at several locations around the Indian Ocean: these appear to be associated with a recently-discovered submarine eighteen-mile wide crater roughly halfway between India and Madagascar.

Similar discoveries have caused some archaeologists to suggest that multiple impacts around four thousand years ago were responsible for widespread flooding around the globe, engendering the many legends mentioned above and terminating several prehistoric civilisations.

Just how many of these significant cometary impacts there have been and when the last one occurred are conjectural: my personal belief is that, directly or indirectly, they have been the cause of the majority of global extinctions and many of mankind's cultural changes.

Some authorities claim to have identified fifty million year cycle of such events: if that's the case, it's worth recalling that the last generally accepted one was sixty five million years ago: the next is fifteen million years overdue!

Other recent research, however, suggests that cometary impacts may be much more frequent – and potentially more devastating – than is generally accepted.

If this is the case, then there is at least one glimmer of cheer: if it were detected early enough, it would be a lot easier to destroy a cometary nucleus than an inbound asteroid . . .

Appendix 1

MODELS OF THE UNIVERSE

MANY of the cleverest people on the planet don't really understand the mathematics behind some of the suggestions that have been made about the origin of the Universe: I'm certainly not in that category, so I shall restrict myself to a brief resumé of each of some of the more widely-discussed theories!

The Steady State Universe

First proposed by Sir James Jeans early in the 20th Century, this model proposes that the Universe is continuously expanding, but maintains a constant mean density. This is achieved by the production of new stars and galaxies at the same rate that existing ones move beyond our ability to observe them.

The Big Bang model

This, the currently-held theory, holds that the Universe is accelerating away from a central super-dense 'singularity'. (It has been observed that the further from us a galaxy is, the faster it is receding) In this model space/time and matter/energy 'decoupled' at the same instant, generally agreed to have been around 13.8 billion years ago.

The Eternal Inflation Multiverse

In this model, the Universe is comprised of a large (infinite?) number of component universes made up of 'normal' non-inflationary space, surrounded by regions undergoing expansion. While these multiverses are continually using up their energy ('heat death') other, new ones are being produced in inflationary regions, so that the Universe as a whole continues forever.

The Oscillating Universe

Basically, this is a development of the Big Bang Theory that attempts to answer the question 'What happens when the Universe runs out of energy?' The suggestion is that the expansion of the Universe will eventually cease, at which time gravity will accelerate all matter back to a central point: this new singularity will then undergo another Big Bang. Theoretically, since the Universe is a sealed system, this process can be repeated indefinitely!

The Projected Flat Hologram Hypothesis

The intriguing theory first proposed in 1997 by physicist Juan Maldacena suggests that our Universe is, in reality, a holographic projection of a two-dimensional image. The PFHH arose from the need to describe gravity within a three-dimensional model of the cosmos, while understanding the behaviour of quantum particles in two spatial dimensions.

The Digital Simulation

Recent astrophysical research suggests that the Universe may have a digital rather than analog mathematical basis. This has led to conjecture that our entire existence is part of a computer simulation run by an alien intelligence.

This concept is by no means brand-new: it reflects the ancient philosophical question: 'Am I dreaming you or are you dreaming me?'

It is the theme of the thought-provoking 1964 novel *Counterfeit World* by American writer Daniel F Galouye. (If you haven't read it, you really should, IMHO!) and of the *Matrix* series of films.

Actually, this suggestion is gaining more and more support among scientific philosophers, not the least because it is the only theory so far that explains the curious observed phenomenon that mankind continually seems to be creating new versions of the observed Universe!

A couple of examples of this:

- Within a few years of a handful of astrophysicists postulating the existence of Black Holes, evidence of their reality was found to be available already in data and photo-banks!

- Until a few years ago, few astronomers took seriously the belief that exoplanets – planets orbiting distant stars – would ever be discovered. Following increasing acceptance by the scientific community that they should be an inevitable result of stellar condensation, new ones are being identified almost weekly!

- In 1943, Thomas Watson (then President of IBM) stated

- "I think there is a world market for maybe five computers."

- In the 1980s, there were very few computers in private hands: many IT specialists could not envision a general need for more than a few hundred. Then a few visionaries began to suggest applications that redefined the possibilities. Thirty years later, it's hard to see how we managed without PCs, laptops and tablets!

Appendix 2

THE 'NUCLEAR WINTER' CONCEPT

HAVING been born just after the Second World War, I grew up in real and constant fear of thermonuclear war.

For much of second half of the twentieth century, the world was divided into two camps: the Soviet Union and the United States and their allies within the Warsaw Pact and NATO.

By the 1960s the two major powers had achieved 'MAD': mutually assured destruction. Put simply: if war had broken out (and it nearly did, on literally hundreds of occasions!) the entire planet would have been devastated: the only surviving humans would have been those who remained underground in self-contained hardened shelters for years. As early as 1948, Albert Einstein put the whole thing into perspective:

"I don't know what weapons World War III will be fought with, but World War IV will be fought with sticks and stones"

It would be very difficult for a modern teenager to grasp the extent to which the lives of children were dominated by the threat of a seemingly unstoppable apocalypse. In my primary school in the East End of London, we practised 'duck and cover' drills every few weeks: the teacher would simulate the sound of an air raid siren, whereupon we would dive under our desks, grab our knees and roll into balls. Quite how this was supposed to protect us from the blast of a twenty megaton H-bomb five miles away in central London was unclear!

During particularly tense episodes of the Cold War, a booklet called *Protect and Survive* would have been issued: this was intended to tell us all how to make our homes safer from the effects of a nuclear attack. Following public clamour, the booklet was eventually made available from HMSO in the late 1970s. It was both chilling and laughable at the same time, advocating such 'emergency strategies' as removing doors and propping them up to form a place of refuge, storing plastic sheets in which to wrap bodies 'until help arrives' and taping around external doors and windows to keep out fallout. The full horror of this post-apocalyptic world was used as the theme of Raymond 'The Snowman' Brigg's book and film *When the Wind Blows*.

Personally speaking, even as a young teenager, the idea of a four-minute warning seemed laughable to me: what was one supposed to do? Boil an egg?

To many people, the worst possible scenario would have been to have *survived* a nuclear war: not just because of the inevitable devastation, hunger, plague and breakdown of society, but more frighteningly, the 'nuclear winter' that was predicted to follow.

Had NATO and the Warsaw Pact exploded their entire nuclear arsenals, (estimated at around 70,000 warheads by 1970) in a single unrestricted exchange, one of the predicted results would have been thousands of tonnes of debris being blasted high into the atmosphere. Research at the time indicated that the resulting soot and dust layer would have reduced incoming sunlight sufficiently to compromise severely photosynthesis by green plants and significantly lower atmospheric temperatures. Globally, many more people would have died as a result than would have been killed by the initial blasts.

When the Alvarez Team first announced their theories about an extraterrestrial impact being responsible for the K-T extinction, they were able to refer to the Nuclear Winter concept as a contributory factor: now the term is widely used, despite its sinister origins not being fully grasped by younger generations.

Appendix 3

THE BIOLOGY OF EVOLUTION

EVOLUTION, the generally agreed mechanism by which existing species gradually evolve into new ones, is a complex process which I'll try to explain in everyday language! The main contributory factors are:

- **Gene mutation / variation**
- **Fitness for survival**
- **Environmental / climatic change**
- **Population isolation**
- **Extinction events**

Let's work our way through these one at a time:

Gene mutation / variation

Each new generation of animal or plant inherits physical and behavioural characteristics from the previous generation, through either **sexual** or **asexual reproduction.** Each new organism must receive a specific 'blueprint' to ensure that it is capable of carrying out the metabolic processes that characterise life (nutrition, growth, sensitivity, movement and so on) and that it develops as a perfect replica of its parent(s)

It receives this data stored in complex arrangements of organic molecules called nucleic acids. In most cases these are contained within the cell in a structure known as the **nucleus**, on dark rod-like structures known as **chromosomes**: discrete regions of these carry 'chemical instructions' that control one or more individual functions: these are called **genes**.

The least complicated method of reproduction is for an organism simply to fragment into two or more new individuals which grow until they are large enough to repeat the process: the familiar Amoeba is the best-known example of this, while Strawberry plants are a higher life-form that can reproduce asexually.

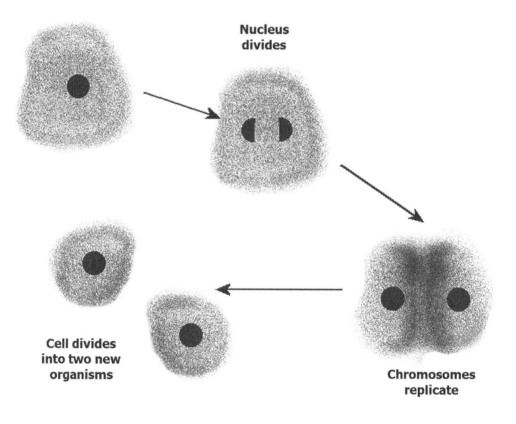

Nucleus divides

Cell divides into two new organisms

Chromosomes replicate

Mitosis in a unicellular organism

Of course, this possibility is not available to animals with high levels of organisation: if a cat were split into two, both halves (lacking the necessary life-support structures) would die.

This form of cell reproduction is how complex organisms like humans grow and repair damaged tissue: the essential feature being the passing on of the parent-cell's full genetic 'blueprint'.

In sexual reproduction two separate parents are required (on Earth, anyhow!) each of which contributes half of the genetic code, packaged in special reproductive cells called **gametes**. In animals these are sperm from the father and eggs from the mother. In this way, within the nucleus of the **zygote** (fertilised egg cell) are pairs of chromosomes carrying genes for the same aspects of structure and function, one of each pair being contributed by each of the parents. So we might inherit the gene for blue eyes from our mother and that for brown eyes from our father. Generally speaking, one or other gene is **dominant** and 'over-rules' the other, **recessive** gene. (In humans this is usually the gene for brown eyes!)

In this way, every single aspect of the structure, appearance, function and metabolism of every single organism on the planet is regulated by the totally individual and virtually infinite combinations of the nitrogenous bases *adenine, guanine, cytosine and thymine* that make up the mega-molecules of DNA (deoxyribonucleic acid) from which the chromosomes are constructed.

As a boy I was once given a model aircraft kit with an incomplete set of plans. Unsurprisingly, the end product when I'd finished making it, looked a bit like the picture on the box-lid in most respects, but differed from it in others. This is the principle behind gene mutation and variation. (Sort of!)

If a gene, or cluster of genes, on a chromosome of, say, an egg cell, is damaged in some way, on fertilisation the resulting zygote will differ from 'normal'. If the mutations are major, the zygote might well fail to develop:

if they are insignificant, however, these differences will continue to be passed on, especially if they make the new indivual 'fitter' to survive. This is the mechanism by which a long series of small changes will, it is generally thought, produce a new species.

There are a number of ways in which an individual's genetic code can be damaged or altered:

Chemical action: It has been shown that certain chemicals may cause abnormal gametogenesis (sex cell production). This may enter the body in food, water or inspired gases.

Radiation: Exposure to high levels of electromagnetic radiation such as gamma, beta, ultra-violet and X-radiation can cause gene mutation: most people are aware that prolonged exposure to sunlight can cause skin cancer as a result of faulty mitosis (growth division). In the same way, radiation can affect gametogenesis, resulting in gene defects: this was tragically demonstrated in the aftermath of the use of nuclear weapons at Hiroshima and Nagasaki at the end of the Second World War.

Old age: As we grow older, transcription – copying of the DNA code during cell replication – may become less accurate. It has been demonstrated, for example, that older mothers are more likely to bear children with the extra 21st chromosome that is associated with Down's Syndrome.

Fitness for survival

The word 'fit' has lots of meanings, but just one in connection with evolution. In simple terms, an organism is biologically *fit* if its inherited characteristics give it a greater propensity to survive long enough to pass its genes on: the more offspring it produces, the greater its biological fitness. (Of course, in many organisms this definition might well also include elements of physical well-being and strength that allow the organism to compete more successfully for a mate.)

Environmental and climatic change

Even quite gradual changes in an environment can have a dramatic impact on the organisms that inhabit it. For example, the drying out of East Africa (possibly caused by India blocking the Monsoon as a result of its tectonic drift northwards) has turned large areas that were once forest into grassland or even desert. Rock shelters in the Sahara depict forest animals such as Antelope, Giraffe and Okapi that lived there just ten thousand years ago. Closer to home, all British bird watchers are aware of the dramatic change in the fortunes of species such as Red-backed Shrike and Wryneck, the virtual extinction of which is not fully understood but seems to be connected with environmental change. But equally, if an organism can adapt rapidly enough, such changes can render it 'fitter' and kick-start an expansion of its population and range: the Mammoth and Woolly Rhinoceros are examples of this. Another is the Peppered Moth, the dark form of which became more abundant during the industrial revolution because it was better camouflaged against soot-covered tree bark than the regular pale form.

Peppered moths

Population isolation

Because of reduced competition and predation, isolated populations of plant and animal often evolve into completely new forms: an example of this would be the various species of Giant Tortoise found on remote islands in the Pacific and Indian Oceans.

Isolation was also a factor in the continued existence and evolution of the monotremes (egg laying mammals) and marsupials (pouched mammals) of Australasia. Since the continent separated from other land masses before the evolution of the highly successful placental mammals, both groups survived the competition that wiped them out everywhere else (with a couple of exceptions!)

Extinction events

In addition to the global extinctions we have discussed earlier, there have been occasions when the decline or demise of one species has allowed another to flourish and evolve at a faster rate, either because of reduced competition, or because of significant environmental changes.

I readily concede that the above is an incomplete and **very** brief account of the mechanisms of genetics and evolution: to the many of you with a far greater depth of knowledge than I possess, I trust you'll understand the need to keep this section accessible to **anyone** who might read it!

Appendix 4

EVOLUTIONARY MIMICRY

AS was discussed in Chapter 3, there are a large number of animal and plant species that strongly resemble completely unrelated forms. In some cases it is not difficult to explain this in Darwinian terms: some others are more perplexing.

The Texas Coral Snake and the Mexican Milk Snake are extremely similar in appearance: predators 'know' to avoid the former, thus benefitting the latter. One might wonder how a predator would learn to avoid the fatal bite of such an organism, or how it could survive to pass this knowledge on to

Coral snake and Milk snake

its descendants! The conventional answer (known as Mertensian Mimicry) suggests that in such cases it is the harmless species that is being mimicked by the dangerous one: but if that were so, what possible advantage would the Milk Snake gain from evolving such a garish colouration?

A curious and beautiful little creature that occasionally visits the herbaceous borders in my garden is the Hummingbird Hawkmoth. Its name reflects the moth's uncanny resemblance to a Humming Bird: it hovers in front of a nectar-rich flower in order to feed, using its long, extensible proboscis. The similarity is heightened by the appearance of the moth's 'tail', which really does look like the spread caudal feathers of a bird. Even more amazing is the

Hummingbird Hawkmoth

insect's eye: instead of the familiar multi-faceted compound eye of a typical moth, this (and several similarly species) have acquired pigmentation that allows their eyes to apparently possess an iris and pupil! The conventional wisdom is that this is an example of **convergent evolution**: the similarity in appearance of the two types is entirely caused by them sharing the same behaviour and feeding niche. (As is the case with sharks and dolphins, for example) But this in no way explains how or why the moth's eye has acquired a false pupil or why a European moth species should closely resemble a species of bird that does not occur over much of its range!

If you're ever lucky enough to see a Hummingbird Hawkmoth, you will be astonished by how closely it resembles a small bird!

The degree to which some orchids have evolved flowers that resemble animals is equally intriguing.

Bee Orchid & Lizard Orchid

Appendix 5

CLASSIFICATION: THE SCIENCE OF TAXONOMY

IN the early days of natural history, animals and plants tended to be put into groups with others that they closely resembled or which shared their method of moving. For example, whales and dolphins were often considered to be fish and Giant Pandas were thought to be raccoons.

Today, classification is based upon the evolutionary and genetic links between organisms: as much as a Rock Hyrax looks a bit like a Guinea Pig, it is known to be a close relative of the Elephants and is included with them (and Manatees and Dugongs) in the taxon Paenungulata!

The modern science of **taxonomy** is based upon the work of the eighteenth century Swedish botanist, physician, and zoologist Carl Linné. His major innovation was the binomial system of classification, whereby every unique organism is assigned a generic name and a specific name. For example, modern humans belong to the genus 'Homo' (always with a capital initial letter!) and the species 'sapiens': thus we are **Homo sapiens**, whereas the extinct species Neanderthal Man has the binomial appellation **Homo neanderthaliensis**. The specific name often refers to an organism's habitat, physical characteristic or may be the name of the first person to identify it as being new to science. (For example: Père David's deer (*Elaphurus davidianus*).

Today, all living things are assigned a place in the 'family tree' of life on Earth, as follows:

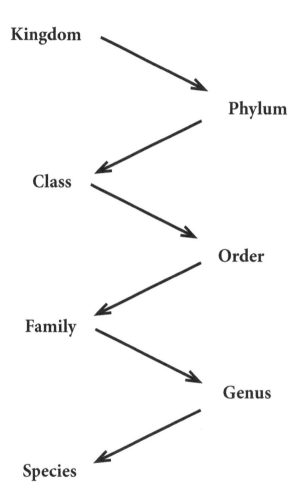

Some of these groups are further divided, so that human beings are classified as follows:

Kingdom	Animalia
Phylum	Chordata
Subphylum	Vertebrata
Class	Mammalia
Subclass	Theria
Infraclass	Eutheria
Order	Primata
Suborder	Anthropoidea
Superfamily	Hominoidea
Family	Hominidae
Genus	Homo
Species	sapiens

... while the European Daisy is classified thus:

Kingdom	Plantae	Plants
Subkingdom	Viridiplantae	
Infrakingdom	Streptophyta	Land plants
Superdivision	Embryophyta	
Division	Tracheophyta	Vascular plants
Subdivision	Spermatophytina	Seed-bearers
Class	Magnoliopsida	
Superorder	Asteranae	
Order	Asterales	
Family	Asteraceae	Sunflowers
Genus	Bellis	
Species	perennis	Daisy

Latin is used in this system because it is a 'dead' language and therefore acceptable globally, because it has always been the language of science and erudition and because it is understood in most countries on Earth.

Appendix 6

METEORITES AND THEIR ORIGINS

4.56 BILLION years ago, probably as a result of pressure fronts moving through the solar nebula (the cloud of dust and gas that surrounded the youthful Sun) particles accreted to form small spherical objects known as **chondrules**.

These grew to form increasingly large chunks of material (chondritic meteoroids) and, eventually, planetissimals: small planets that had differentiated to form a nickel iron core with outer mantle and crust layers. Collisions between these produced the eight planets and asteroids of the present Solar System, but large amounts of ancient, unassociated material remained in interplanetary space.

When smallish bits – say the size of a grain of rice – run into the Earth, they become very hot because of friction with the air they are passing through: then we might see a 'shooting star' or **meteor**, to give it its proper name. From time to time someone might be lucky enough to find a piece of rock – or metal – from Space that was large enough to survive the passage through the atmosphere and reach the ground: these objects are called meteorites.

The most abundant types of meteorite are:

Stony meteorites. These are left over from the formation of the Solar System or come from the surface of other planets. A couple of hundred meteorites (known as achondrites) are known to have been blasted from the surface of the Moon or Mars by meteorite or cometary impacts: about the same number originated in the same way on asteroids such as 4-Vesta.

Stony irons. Pallasites and mesosiderites are thought to derive from the core-mantle layer of shattered planets, or from collisions between stony and nickel-iron bodies.

Iron meteorites These may have either condensed directly within the Solar Nebula or be the cores of disrupted, differentiated planetissimals.

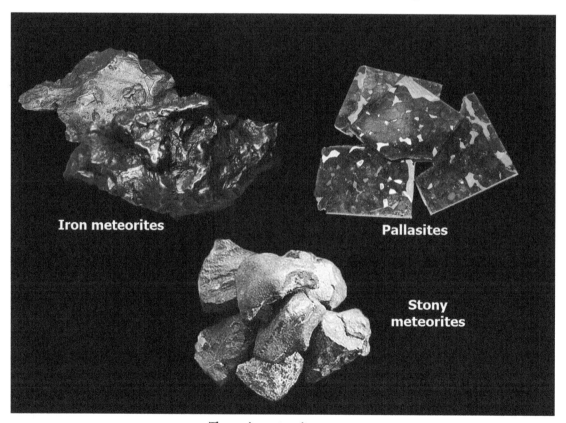

Iron meteorites

Pallasites

Stony meteorites

The main meteorite groups

It's worth reflecting on the fact that 300 tonnes of meteoritic material lands on Earth each day, so there really is a lot of it out there: if you know what you're looking for and can carry out a few simple tests, you might well be lucky enough to find one!

Recognising a meteorite

The first thing to stress about meteorites is that the great majority are actually quite ordinary-looking: the examples shown in the few books on the subject are generally chosen from thousands just because they are particularly attractive!

If you find a glittery, beautifully-marked or shaped rock it **won't** be a meteorite: here are five ways to help you recognise a real one!

- **Is it attracted to a magnet?** The vast majority of stony and iron meteorites are strongly attracted to a decent magnet because of the nickel-iron they contain. Only the very rarest achondrites are not.

- **Does it weigh more than it looks as if it should?** For the same reason as above, meteorites are generally surprisingly dense. Common chondrites are usually in the density range of $3.0 - 3.7$ g/cm^3

- **Does it have a fusion crust?** A meteorite acquires a thin, black crust because of the high temperatures generated by friction as it passes through the atmosphere. This will fairly quickly disappear on exposure to wind and rain, but a fresh fall should have a sooty, matt surface

- **Does it display flowlines and/or regmaglypts?** The frictional forces mentioned above often sculpt the surface of a meteorite, producing thumbprint-like depressions called regmaglypts. If the meteorite has orientated itself in flight, it might also show shallow flowlines where molten rock has streamed away from the hottest front surface

- **Does it contain nickel?** Although nickel-containing minerals are not rare on Earth, most rocks wouldn't give a positive response to a nickel test. If you want to test your own finds, you can buy a test kit for less than £10 from a chemist's shop.

Appendix 7

THE MOON AND ITS CRATERS

THE Earth, as I mentioned earlier, is not the only body in the Solar System to display evidence of major impact events: every object with a solid surface is pock-marked with craters.

The most familiar of these is, of course, the Moon: even a pair of binoculars will reveal a couple of dozen of the larger craters. The photographs below were taken with an inexpensive DSLR and a 500mm telephoto lens and show literally hundreds of craters.

Perhaps unsurprisingly, until comparatively recently these were believed to have been volcanic in origin. The central peaks displayed by many were identified as volcanic cones, while the often-distant crater rims were considered to be ejecta, flung much further than on Earth as a result of the Moon's reduced gravity (around one sixth of the Earth's.)

Today, it is generally held that following an initial heavy bombardment phase that created the crater-fields covering much of the Moon's surface, several dozen much larger impacts were responsible for the vast circular basalt *maria basins*. ('Maria' means 'seas' in latin: it was believed by early astronomers that these darker regions bodies of water which is reflected in

Craters on the Moon

their names: Sea of Tranquillity, Ocean of Storms, Bay of Rainbows and so on!) These basalt flood plains also bear the marks of subsequent impacts, some of which may well have occurred comparatively recently. The large ray crater, Tycho, for example, is considered by some astronomers to have been formed just a million or so years ago: certainly its rays of pale ejecta are superimposed over much of the lunar surface.

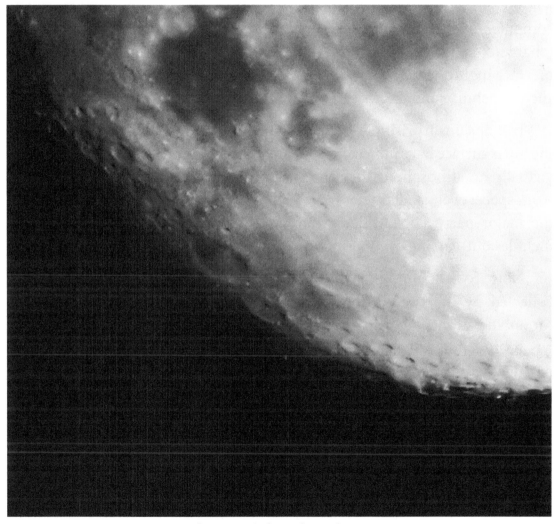

The crater Tycho and its pale rays

When I was a teenager, I made my first 'proper' telescope (a four-inch reflector) from the cardboard tube that had once had a carpet wrapped around it and a mirror I made myself with a grinding kit I bought from a Scottish supply company. It sat upon a simple equatorial mount that had to be rotated by hand to keep an object under observation in the centre of the field of view. These days things are a lot different: a few hundred pounds will buy an optically superb 20cm Newtonian reflector on a motor-driven, computer steered 'go to' drive. This can be fitted with an inexpensive CCD (charge-coupled device) that would allow even inexperienced amateurs to produce high-quality images of the Moon, planets and even deep-sky objects such as nebulae and galaxies.

Perhaps unsurprisingly, therefore, small, apparently meteoric impacts on the lunar surface have increasingly been observed – and even videoed. Having virtually no atmosphere, the Moon is exposed to a constant bombardment from space: even small meteoroids will impact upon its surface. These form the major part of the wider group of **Transient Lunar Phenomena** that includes out-gassing events and possibly electro-static discharges.

Now if we assume that the three hundred or so kilos of rock samples collected by the twelve Apollo Astronauts did actually originate on the Moon, the strange fact is that virtually none of the material is meteoric! It has been claimed that a few tiny fragments of meteorite were discovered during investigation of the Apollo moon rocks (currently stored at Lyndon B. Johnson Space Centre in Houston, Texas, and at the White Sands Test Facility, Las Cruces, New Mexico), while other items of interest to meteoricists are two lumps, catalogued as A12013 and A14425, collected on the Apollo 12 mission. These apparently contain fragments of black tektite (the glassy impactite mentioned earlier) suggesting either that tektites may have originated on the Moon or that debris from major impacts on the Earth has been ejected with such force that it escaped our gravity and travelled 400,000 km across space.

Given the lack of an atmosphere or weather systems, shouldn't the many millions of impacts that created the Moon's landscape have left the surface littered with chunks of meteorite? Not just the alleged spherical, pink regolith particles, but proper, big pieces scattered over the whole planet. The Mars rover 'Opportunity' (with its restricted freedom of activity and limited decision-making competence) managed to find half a dozen: why didn't the Apollo EVAs enjoy similar success?

To me the only likely answer (other than that dealt with by my earlier book *Our Forbidden Moon*) is that the majority of the cratering on the lunar surface – and certainly the vast maria basins – were the result of cometary impacts. So what happened to all the ice? Well: most of it would have sublimed directly into space as a gas. Given that the surface temperature of the Moon rises to over a hundred degrees Celsius during the daytime and that the Moon's gravity is just a fraction of the Earth's, the majority of any ice or water remaining after a cometary impact would quickly boil off. Having said which, it has been claimed that water ice has been detected beneath the surface of the polar regions: these, of course, as is the case here on Earth, are much colder than the rest of the Moon.

IMAGE CREDITS

All images used in this book are either the author's own, or are in the public domain. While every effort has been made to discover the original publisher of an image and give credit, this has not always been possible.

Image 1   Comet Pan-starrs *(David Bryant)*

Image 2   The author and Rusty Schweickart *(David Bryant)*

Image 3   Meteorite, polished to show chondrules *(David Bryant)*

Image 4   Alfred Wegener *(Unknown – Bildarchiv Foto Marburg)*

Image 5   Subduction at a plate margin *(David Bryant)*

Image 6   Geological Periods *(David Bryant)*

Image 7   My first fossils! *(David Bryant)*

Image 8   Baron Georges Cuvier *(Public domain)*

Image 9   Micro-tektites from Hell Creek *(David Bryant)*

Image 10   Sooty K-T material from Alberta *(David Bryant)*

Image 11   Possible formation of flood basalts *(David Bryant)*

Image 12   Some large terrestrial craters *(Wikipedia Commons)*

Image 13   Dawn arrival of a meteoroid *(David Bryant)*

Image 14   Damage caused by variously-sized impactors *(David Bryant)*

GLOSSARY

Ionosphere: a layer of ionised gases forming one of the outer layers of the Earth's atmosphere at an altitude of 60km – 1,000km

Van Allen Belts: two regions of high-energy particles held in position 1,000km – 60,000km above the equator by the Earth's magnetic field.

B612 Foundation: an organisation that promotes the tracking of asteroids and investigates methods of deflecting or destroying any that may pose a threat to life on Earth. The group derives its name from the home-asteroid of the hero of *The Little Prince* by Antoine de Saint-Exupéry.

Nuclear fusion: the combination of two or more atomic nuclei to form a larger one. When this takes place between nuclei lighter than iron, energy is generally released: hydrogen fusion is the basic energy-generation process within stars.

Electrostatic attraction: this is the force that draws together objects or particles with opposite electric charges. It is the force that causes a balloon to stick to the ceiling if it is first rubbed against a piece of wool.

Similar charges repel each other.

Equilibration / differentiation: the process undergone by planetary bodies (and their component rocks) whereby heavier atoms and their compounds sink inwards to form a dense core: the primitive structure of the planet and its minerals are destroyed.

Hydrosphere: the thin layer of water that makes up the seas, oceans, lakes and rivers upon the surface of a planet like the Earth.

Mass: a measurement of the amount of matter within an object. This doesn't change, regardless of the objects position, whether within a gravitational field or in free-fall. Its units are grams and kilograms

Weight: if an object with mass is subject to gravitational attraction or another external acceleration, it acquires the property of force:

$$P = m \times a$$

(force = mass x acceleration.) A mass of one kilogram on Earth is accelerated by gravity at 9.81 metres per second, per second, giving it a downwards force (or weight) of 9.81 Newtons. To give an example: a spacecraft standing on the surface of the Moon is subject to a much smaller gravitational acceleration (about a sixth of the Earth's.). It therefore has one sixth of the weight that it would have on the Earth's surface, although it has the same mass as it would on Earth.

Cartographer: a map-maker

Tessellating: fitting together like a jigsaw puzzle.

Flora and Fauna: plants and animals

Mantle: the mineral layer that surrounds the core of a rocky planet.

Hypersaline: a body of water containing very large amounts of dissolved salts, particularly sodium chloride ($NaCl$) and sodium hydrogen carbonate ($NaHCO_3$).

Gamete: a reproductive cell that contains half the normal amount of nucleic acid. In animals the male gametes are called *sperm*, in flowering plants *pollen*: the female gametes in both cases are known as *ova*. It is usually the smaller male gamete that moves towards the female one, either by being self-motile or by being carried by water or the wind.

Zygote: the new cell formed when a male gamete *fertilises* a female gamete. This possesses the normal amount of nucleic acid / chromosomes, but has acquired this from two different individuals.

Cephalopods: a group of highly successful molluscs that have existed on Earth since the late Cambrian or early Ordovician Periods. Their main features are numerous arms and / or tentacles, a parrot-like beak and the ability to propel themselves backwards using water projected from a siphon. Modern forms include Squid, Cuttlefish and Octopuses.

Herbivore: an animal that subsists on a diet of plants.

Carnivore: an animal (or rarely plant!) that subsists by eating animals

Omnivore: an animal that eats both plants and animals (eg most humans and bears)

Carrion-eater: this is an organism that subsists on a diet of dead and often decaying animal tissue.

Sauropod: from the Greek for 'lizard foot', these were one of the major groups of dinosaurs and included the familiar giant forms such as the genera *Brachiosaurus*, *Diplodocus*, and *Brontosaurus*

Theropod: again, from the Greek, this time meaning 'beast foot'. This group included the familiar carnivorous dinosaurs Tyrannosaurus, Allosaurus and Spinosaurus, as well as smaller types that evolved into the birds.

Palaeontology: the study of ancient life-forms.

Aerosols: in the context of post-impact climate change, these are colloids of fine particles or droplets of liquid suspended in the atmosphere.

Hibernation: surviving the winter by entering a state of torpor until the spring.

Aestivation: surviving the summer or drought conditions by entering a state of torpor until the autumn.

Iconoclast: one who repudiates or disdains conventional or traditional beliefs. Originally, this referred to the physical destruction of religious symbols or images.

Anthropogenic: 'produced by mankind': in relation to climate change, the widely-held but still questionable belief that it is the emission of so-called 'greenhouse gases' such as carbon dioxide and methane by humans and their livestock that is responsible for raising the temperature of the atmosphere. Many reputable climatologists have questioned the motives of politicians in promoting this idea and have frequently seen their careers collapse as a result.

Widmanstätten pattern: many nickel-iron meteorites display a characteristic crystalline structure when cut, polished and etched. The size of the crystals is so large that they must have required thousands of years of growth following the planetary disruption that released them.

Regolith: the outer 'soil' of a planetary or cometary body.

Ion: a charged atom. Negative ions have extra electrons, whiloe positive ions are atoms that have lost one or more electrons. Ionic bonding occurs when a positive ion is electrostatically joined to a negative one. (ie: Na+ and Cl- bond to form NaCL, sodium chloride.)

Aperiodic: in astronomy, an object possessing a trajectory that causes it to enter the Solar System on a parabolic path that prevents it returning. Periodic comets are in elliptical orbits that allow predictions of their regular apparitions.

Isotopes: different forms of the same atom that contain differing numbers of neutrons in their nuclei. Hydrogen atoms, for example, may have none, one or two neutrons. Isotopes of the same element generally have more or less identical *chemical* properties but different *physical* properties.

Mass Number: the quantity of protons in the nucleus of an atom: this is the same as the number of electrons that exist in shells around the nucleus. Hydrogen has an atomic number of one, Carbon six, Uranium ninety-two and so on.

Atomic mass: this is the average mass of an atom of any element, taking into account the various isotopes of that element. It is the total of all the particles in the nucleus (protons and neutrons)

Iridium: a dense silver-white metal similar to Platinum, with a mass number of 77. It is rare on Earth but much more abundant in cometary regoliths and meteorites. Its presence in terrestrial strata is an indicator of an impact event.

Kinetic energy: energy possessed by a moving object. When such an object comes to an abrupt stop, the energy is transformed into heat, light and sound.

Megaton: one million tons. In relation to the explosive output of a nuclear device, a one megaton atom bomb has the explosive power of a million tons of TNT (trinitrotoluene, $C_6H_2(NO_2)_3CH_3$)

Epicentre: the point of origin of an explosion: 'ground zero'.

Astrobleme: from the Greek 'astron' and 'blema', meaning "star wound", an alternative name for an impact crater.

BIBLIOGRAPHY

Some books that you might enjoy!

Astronomy

Big Bang: the Origin of the Universe Simon Singh

Evolution from Space Sir Fred Hoyle & Chandra Wickramasinghe

Rocks from Space O. Richard Norton

Field Guide to Meteors and Meteorites O. Richard Norton

Atlas of Meteorites Monica Grady, Giovanni Pratesi & Vanni Moggi

Atlas of Great Comets Ronald Stoyan & Storm Dunlop

Comet Carl Sagan & Ann Druyan

Comets: Visitors from Deep Space David J Eicher

Prehistoric life and evolution

The Origin of Species Charles Darwin

Georges Cuvier, Fossil Bones, & Geological Catastrophes Martin J S Rudwick

Basic Palaeontology Michael J. Benton & David A.T. Harper

The Great Extinctions Norman MacCleod

Geology and tectonics

Global Tectonics Philip Kearey, Keith A. Klepeis & Frederick J. Vine

Earth Structure Ben A. van der Pluijm & Stephen Marshak

DISCLAIMER

David Bryant while lecturing at Stargazing Live

SPACEROCKS UK

DAVID BRYANT, BSc, Cert Ed is the only full-time meteorite dealer in England. His company 'Spacerocks UK' holds a complete inventory of all meteorite types, from 4.5 billion year old common chondrites, to iron meteorites, rare and beautiful pallasites and even pieces of the Moon, Mars and the Asteroid Vesta!

He is a member of the IMCA, (International Meteorite Collectors Association) and all his items are sold with an A4 factsheet and guarantee of authenticity. David has delivered lectures about meteorites at meetings of the British Astronomical Association, the Society for Popular Astronomy, for the BBC's 'Stargazing Live' events and astronomical societies all over the country.

David sells his meteorites at rock and mineral shows around the UK, from Cambridgeshire to Devon and at most of his public lectures.

All these items can be purchased by 'phone, 01603 715933 or from the SPACEROCKS UK website at:

http://www.spacerocksuk.com
email: info@spacerocksuk.com

His wife, Linda, makes a wide range of beautiful meteorite and impactite jewellery using solid silver chains and findings, which are available from her website:

http://www.space-jewellery.co.uk

David and Linda Bryant seen here recently, displaying items available for purchase
at one of the Spacerocks UK stalls

OUR FORBIDDEN MOON

By David Bryant *(Foreword by Nick Pope)*

140 pages, some in colour

David Bryant's book *Our Forbidden Moon* has taken fifteen years of meticulous planning and research to write. During encounters with over thirty astronauts and cosmonauts, including seven of the twelve alleged Moonwalkers, the author gradually became aware of a number of major inconsistencies in their recollections of the Apollo program. Furthermore, in occasional unguarded moments, several space travellers have revealed personal experiences of the UFO phenomenon and hinted at even more dramatic events. *Our Forbidden Moon* examines these revelations and considers whether there might be a link between UFOs, extraterrestrial races and mankind's forty-year failure to travel beyond low Earth orbit. The author uses his knowledge gained during forty years as a teacher, lecturer and respected authority on spaceflight and meteoritics to ask controversial questions and provide convincing solutions.

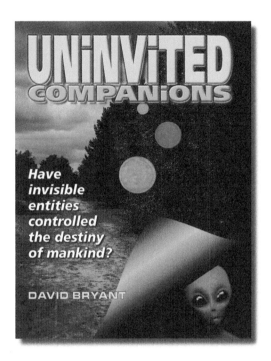

UNINVITED COMPANIONS

By David Bryant *(Foreword by Lionel Fanthorpe)*

168 pages, some in colour

Uninvited Companions is an examination of the paranormal phenomena known as orbs: the strange spheres of light that frequently appear on video and still photographs or are even, at times, seen by the unaided eye. With a BSc in Biological Sciences and Astronomy and a lifetime as a teacher and lecturer, author David Bryant has reached some new and startling conclusions about these contentious objects! The book examines other, apparently unrelated paranormal happenings, and consider whether they might be linked in some way to the orb phenomenon.

Lightning Source UK Ltd.
Milton Keynes UK
UKOW07f1345240616

276997UK00005B/23/P